QUANTUM
PHYSICS FOR BEGINNERS

Learn the Secrets of Quantum Mechanics, Understand Essential Theories Like the Theory of Relativity, and the Entanglement Theory, and Exploit the Law of Attraction

Quantum Physics for Beginners

© Copyright 2021 by (DANIEL LONG) —
All rights reserved.

This document is geared towards providing exact and reliable information regarding the topic and issue covered. The publication is sold with the idea that the publisher is not required to render accounting, officially permitted or otherwise qualified services. If advice is necessary; legal, or professional, an experienced individual in the profession should be ordered. - From a Declaration of Principles, which was accepted and approved equally by a Committee of the American Bar Association and a Committee of Publishers and Associations.

In no way is it legal to reproduce, duplicate, or transmit any part of this document in either electronic means or printed format. Recording of this publication is strictly prohibited, and any storage of this document is not allowed unless with written permission from the publisher. All rights reserved.

The information provided herein is stated to be truthful and consistent in that any liability, in terms of inattention or otherwise, by any usage or abuse of any policies, processes, or directions contained within is the solitary and utter responsibility of the recipient reader. Under no circumstances will any legal responsibility or blame be held against the publisher for reparation, damages, or monetary loss due to the information herein, either directly or indirectly.

Respective authors own all copyrights not held by the publisher.

The information herein is offered for informational purposes solely and is universal as so. The presentation of the information is without a contract or any type of guarantee assurance.

The trademarks used are without any consent, and the publication of the trademark is without permission or backing by the trademark owner. All trademarks and brands within this book are for clarifying purposes only and are owned by the keepers themselves, not affiliated with this document.

TABLE OF CONTENTS

INTRODUCTION 8

CHAPTER 1: THE START OF QUANTUM PHYSICS 14

   1.1 BRIEF HISTORY 17

   1.2 BASIC CONSTANTS IN NATURE 20

CHAPTER 2: THE PRINCIPLE OF WAVE- PARTICLE DUALITY — AN OVERVIEW 22

   2.1 THE PRINCIPLE OF WAVE-PARTICLE DUALITY OF LIGHT 24

   2.2 THE PHOTOELECTRIC EFFECT 24

   2.3 THE COMPTON EFFECT 29

   2.4 THE PRINCIPLE OF WAVE-PARTICLE DUALITY OF MATTER 31

   2.5 ENERGY QUANTIZATION IN MATTER WAVES 32

   2.6 ATOMIC STABILITY UNDER COLLISIONS 34

   2.7 ENERGY SCALES 36

   2.8 THE STABILITY OF ATOMS AND MOLECULES AGAINST EXTERNAL ELECTROMAGNETIC

      RADIATION 41

   2.9 PROBABILISTIC INTERPRETATION OF MATTER WAVES 44

   2.10 QUANTUM JUMPS FROM HIGHER TO THE LOWER ENERGY STATES AND ATOMIC SPECTRA .. 47

CHAPTER 3: CONCEPTS OF QUANTUM PHYSICS 52

   3.1 PLANCK'S RADIATION LAW 52

3.2 Einstein and the Photoelectric Effect 53

3.3 Bohr's Theory of the Atom 54

3.4 Scattering of X-rays 56

3.5 De Broglie's Wave Hypothesis 58

3.6 Basic Concepts and Methods 58

3.7 Schrödinger's Wave Mechanics 59

3.8 Time-Dependent Schrödinger Equation 64

3.9 Electron Spin and Antiparticles 65

3.10 Identical Particles and Multielectron Atoms 67

3.11 Tunneling 69

3.12 Axiomatic Approach 71

3.13 Incompatible Observables 73

3.14 Heisenberg Uncertainty Principle 76

3.15 Quantum Electrodynamics 81

3.16 Quantum Harmonic Oscillator 83

3.17 Quantum Entanglement 86

CHAPTER 4: THE INTERPRETATION OF QUANTUM MECHANICS 98

4.1 The Electron - Wave or Particle? 98

4.2 Hidden Variables 100

4.3 Paradox of Einstein, Podolsky, and Rosen 103

4.4 Measurement in Quantum Mechanics 107

CHAPTER 5: PLANCK'S CONSTANT IN ACTION 112

    5.1 THE INVISIBLE WORLD OF THE ULTRASMALL 113

    5.2 A SLIGHTLY CHANGING CONSTANT 114

CHAPTER 6: APPLICATIONS OF QUANTUM PHYSICS 118

    6.1 DECAY OF THE KAON 118

    6.2 CESIUM CLOCK 121

    6.3 A QUANTUM VOLTAGE STANDARD 125

CONCLUSION 128

Introduction

Teaching quantum mechanics in secondary school has inherent difficulties, such as students' inadequate previous knowledge of mathematics and classical physics, as well as a restricted amount of time. As a result, it's not shocking that less effort has been made to strengthen teaching at this level than at the university's introductory level. Nevertheless, it seems that there is room for improvement. In this article, the naive photon concept is scrutinized, and a new approach is proposed. The photoelectric effect is traditionally considered first with varying degrees of sophistication and elaboration. Experiments show the failure of classical electrodynamics, and it is argued that they imply:

electromagnetic wages –> particles (photons)

A photon has the energy *($W = h\upsilon$)* and momentum *($p = W/c = h\upsilon/c = h/\lambda$)* of a particle with zero rest mass moving with the velocity of light *in vacuo c.* The Planck constant *h* is typical for quantum mechanics, *υ* being the frequency and *$\lambda = c/\upsilon$* the wavelength. Symmetry considerations lead to the de Broglie hypothesis:

matter waves <– *particles* (e.g., electrons)

Matter waves are associated with electrons, and the same relation as for photons is applied to the wavelength, but the momentum is **p** = **mv**, **m** being the mass and **v** the particle velocity. Interference experiments with electrons verify the hypothesis. Later the wave-particle duality of electrons is discussed. Electrons are described either as particles or as waves, both concepts being familiar from classical mechanics. Further arguments based on standing matter waves lead to discrete energy levels and to the time independent Schrodinger equation.

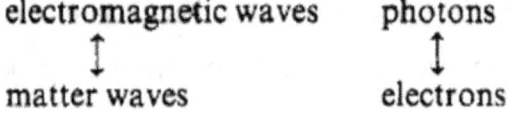

Both parts of the symmetry are; however, open to serious objections. The state function (or the quantum mechanical wavefunction) *ψ(x,t)* describing an electron is not a classical wavefunction *E(x,t)* describing electromagnetic waves (**E** being the electric field). A form of the equation of motion for electrons, the Schrodinger equation:

$$-(\hbar^2/2m)\nabla^2\Psi + V(x)\Psi = i\hbar\partial\Psi/\partial t$$

(**ℏ** = **h/2π** and V(x) is the potential energy), differs from the classical wave equation:

$$\nabla^2 E = c^{-2}\partial^2 E/\partial t^2$$

The first contains the imaginary unit **i** and the first time derivative, while the second is real and contains the second time derivative. The state function is complex, whereas the classical wave function is real. It is true that classical wave functions can be written in the complex form to shorten calculations, but only their real (or imaginary) part is significant. The real function $E(x,t)$ is in principle accessible to measurement. The time dependence of it at a fixed point can be observed on an oscilloscope if the microwave region's frequency is not too high. (The frequency can be directly measured far into the infrared region.) On the contrary, neither the real nor the imaginary part of a state function is directly accessible to measurement, only the norm.

$$\Psi^*\Psi = (\text{Re}\Psi)^2 + (\text{Im}\Psi)^2$$

Being observable in principle.

At this point, it is instructive to mention Landss paradox. If the de Broglie wavelength $\lambda = h/p$ were a true spatial period in a nonrelativistic wave, i.e., a length, it should be invariant under a Galilean transformation: $\lambda' = \lambda$ and thus $p' = p$. At the same time, the momentum of an electron is transformed as $p' = p - mv_o$ if the second inertial reference frame is moving with velocity v_o with respect to the first in the direction of the momentum $p = mv$ of the electron. This contradiction is due to the unjustified demand that the phase of a state function is invariant under a Galilean transformation. Indeed, the phase of a nonrelativistic wavefunction is invariant, but the state function under a Galilean transformation acquires an additional phase factor, and only its norm is invariant.

By demanding form invariance of the Schrodinger equation of a free particle, it can be shown that the appropriate transformation is **h/λ' = h/λ - mv₀** (Levy-Leblond 1976).

Students and teachers and textbook writers sometimes have difficulty distinguishing a state function from a classical wave function. As an example, let us quote the textbooks of Halliday and Resnick. In the 1966 edition of parts I and II, it is asserted on p1207 that the matter wave is described in strict analogy with a classical wave function.

$$y = (2y_m \sin(n\pi x/l))\cos \omega t, \quad n = 1,2,3\ldots$$

By

$$\Psi = (\psi_m \sin(n\pi x/l))\cos \omega t, \quad n = 1,2,3,\ldots$$

In the 1978 edition of part 2 on p1124, the hydrogen ground state is described by

$$\Psi = (\pi a^3)^{-1/2} e^{-r/a} \cos \omega t.$$

Both functions are not solutions to the Schrodinger equation. One could add some further distinctions between a state function and a classical wave function (Rüdiger 1976). But let us, instead, note a real analogy. The time independent Schrodinger equation and the electromagnetic amplitude equation have the same form:

$$\nabla^2 \psi + [2m(W - V(x))/\hbar^2]\psi = 0$$
$$\nabla^2 \zeta + (\omega/c)^2 \zeta = 0.$$

Thereby, $\Psi = \psi \exp(-iW't/h)$ and $E = \zeta \cos \omega t$. A position dependent potential $V(x)$ corresponds to an inhomogeneous medium with position dependent refractive index $n(x)$, since

$$\omega/c = (2\pi/\lambda_0)n(x)$$

where λ_0 is the wavelength in *vacuo*. For bound states, i.e., for boundary conditions of the type (real) solutions of both equations may have the same form.

$$\psi(x \to \pm\infty) = \zeta(x \to \pm\infty) = 0,$$

A photon moving with the velocity of light does not belong to nonrelativistic quantum mechanics, which is based on classical mechanics. In nonrelativistic quantum mechanics, Heisenberg's indeterminacy relation $\delta x \delta p \geq h$ is valid, δp being the indeterminacy of a momentum component and δx the indeterminacy of the corresponding coordinate at a simultaneous measurement of both variables. So, either the position of a particle ($\delta x = 0$, $\delta p \to \infty$) or its momentum ($\delta p = 0$, $\delta x \to \infty$) can be determined exactly. In a relativistic theory, the situation is completely different. The indeterminacy of position is (Landau and Peierls 1931)

$$\delta x \geq hc/W$$

Irrespective of the indeterminacy of momentum, W being the total energy makes photon with $W = h\nu$.

$$\delta x \geq h/p = \lambda.$$

So, speaking of the position of a photon has no meaning outside the domain of geometrical optics. On the other hand, the indeterminacy of the momentum is:

$$\delta p \gtrsim h/c\tau,$$

Irrespective of the indeterminacy of position. So, the momentum can be exactly determined only if the measuring time t is unlimited, i.e., if the relativistic particle is free.

Finally, there is no combination of the electric and magnetic field that would transform as probability density should (Peierls 1979). Thus, a photon cannot be described in a similar way to an electron. A photon cannot be considered a point-like particle. Also, the fuzzy ball picture of the photon can lead to difficulties (Scully and Sargent 1972, Sargent *et ul* 1974). A straightforward introduction of the photon is accomplished only through field quantization in quantum electrodynamics. In this case, the correspondence principle does not lead to a classical limit. The expectation value of the field in a state with a given number of photons vanishes even at the limit of a very large photon number. A classical limit for this expectation value is obtained only for a superposition of photon states with a poisson distribution, known as the coherent state.

CHAPTER 1:

The Start of Quantum Physics

As early as 500 BC, the Greek Philosopher used to debate whether a matter could be split up into an infinite amount of smallest part while preserving the properties of the atom or a matter as a whole. The idea was finally answered in the late 19th century because of the advancement in Statistical Mechanics, Chemistry, and Brownian Motion. A famous scientist named Feynman answered a question of a reporter: if we were to send a single message to alien life in the universe to show the development or advancement on Earth, what would it be? He said: "Matter is composed of atoms." So, it was firmly established that matter is composed of tiny atoms. But what does an atom look like? It was still a myth to resolve for the scientists.

Of course, we now know that atoms are composed of a nucleus and electrons and that the nucleus, in turn, is built from protons and neutrons themselves built out of quarks and gluons. Constructing a viable model for the electronic structure of atoms is what originally and primarily drove the development of quantum mechanics. The existence and stability of atoms are a purely quantum mechanical effect.

Without the Pauli exclusion principle and the shell structure of electrons, we would lose the chemical and physical properties that distinguish different elements in the Mendeleev table, and there would be no chemistry as we know it.

Similarly, the molecular and collective properties of conductors and insulators of heat and electricity have quantum origins as do the semi-conductors used in building transistors and integrated circuits. Atomic and molecular spectral lines, observed from distant stars and nebulae, have allowed astronomers to conclude that the visible matter in the universe at large is the same as that found on Earth. The systematic displacements of these lines inform us of the velocities and the distances of these objects. In summary, quantum physics and quantum phenomena are pervasive in modern science and technology.

Quantum mechanics deals with the behavior of matter and light on the atomic and subatomic scale. It attempts to describe and account for the properties of molecules and atoms and their constituents — electrons, protons, neutrons, and other more esoteric particles such as quarks and gluons. These properties include the interactions of the particles with one another and with electromagnetic radiation (i.e., light, X-rays, and gamma rays).

The behavior of matter and radiation on the atomic scale often seems peculiar, and the consequences of quantum theory are accordingly difficult to understand and believe. Its concepts frequently conflict with common-sense notions derived from observations of the everyday world. There is no reason, however, why the behavior of the atomic world should conform to that of the familiar, large-scale world. It is important to realize that quantum mechanics is a branch of physics and that the business of physics is to describe and account for the way the world — on both the large and the small scale — actually is and not how one imagines it would like it to be.

The study of quantum mechanics is rewarding for several reasons. First, it illustrates the essential methodology of physics. Second, it has been enormously successful in giving correct results in practically every situation to which it has been applied. There is, however, an intriguing paradox. Despite the overwhelming practical success of quantum mechanics, the foundations of the subject contain unresolved problems — in particular, problems concerning the nature of measurement. An essential feature of quantum mechanics is that it is generally impossible, even in principle, to measure a system without disturbing it; the detailed nature of this disturbance and the exact point at which it occurs are obscure and controversial. Thus, quantum mechanics attracted some of the ablest scientists of the 20th century, and they erected what is perhaps the finest intellectual edifice of the period.

1.1 Brief History

Physics, known at the end of the 19th century, falls into three fields:

- Classical mechanics (Hamilton, Newton, Euler, Lagrange, Leverrier...)
- Electromagnetism (Gauss, Faraday, Maxwell, Coulomb, Lorentz, Ampère...)
- Statistical mechanics and thermodynamics (Boltzmann, Joule, Carnot...)

Around that time, Lord Kelvin apparently stated that physics just had a few loose ends to tie up and that physics would be complete by 1900. Nothing about special relativity, which was invented to resolve a conflict between electromagnetism and mechanics, and in doing so, profoundly altered our concepts of space and time.

The most radical revolution, however, was quantum mechanics. The key experiments that were crucial in the development of the quantum theory may be summarized as follows:

- Discrete emission and absorption spectra of simple atoms (Balmer, ...);
- Existence of radioactivity (Becquerel 1896);
- Frequency dependence of back body radiation (Planck 1900);
- Photo-electric effect (Einstein 1905);

- Existence of a hardcore (i.e., the nucleus) inside atoms (Rutherford 1909);
- Incompatibility of the planetary atom with classical electrodynamics (Bohr 1913);
- Discovery of strong and weak forces (Rutherford 1920);
- Electrons diffract just as light does (de Broglie 1923);
- Need for a Pauli exclusion principle in the Bohr model of atoms (Pauli 1925);
- Goudsmit and Uhlenbeck discover spin (1925).

This wealth of novel experimental facts, and their apparent contradiction with (classical mechanics and electromagnetic) theory, led to the formal development of quantum mechanics:

- 1925: Heisenberg introduces matrix mechanics and quantized the harmonic oscillator.
- 1926: Schrödinger invents wave mechanics (inspired by de Broglie's particle/wave duality) and shows the equivalence of matrix and wave mechanics.
- 1927: Dirac gives what is to be the present-day formulation of quantum mechanics.

Subsequent developments primarily include quantum field theory; a theory that unifies quantum mechanics and special relativity. Its foundations were laid in 1928 by Dirac, Heisenberg, and Pauli; but its elaboration continues actively still today. It is important to realize though there has been no serious need since 1927 to alter quantum mechanics' fundamental principles, either for experimental or theoretical reasons. For extensive accounts of quantum mechanics' history and conceptual development, the reader is referred to the books by Jammer and by Pais.

At a fundamental level, both radiation and matter have characteristics of particles and waves. Scientists' gradual recognition that radiation has particle-like properties and that matter has wavelike properties provided the impetus for the development of quantum mechanics. Influenced by Newton, most physicists of the 18th century believed that light consisted of particles called corpuscles. From about 1800, evidence began to accumulate for a wave theory of light. At about this time, Thomas Young showed that if monochromatic light passes through a pair of slits, the two emerging beams interfere so that a fringe pattern of alternately bright and dark bands appears on a screen. The bands are readily explained by a wave theory of light. According to the theory, a bright band is produced when the crests (and troughs) of the waves from the two slits arrive together at the screen; a dark band is produced when the crest of one wave arrives at the same time as the trough of the other, and the effects of the two light beams cancel. In 1815, a series of experiments by Augustin-Jean Fresnel of France and others showed that, when a parallel beam of light passes through a single slit, the emerging beam is no longer parallel but starts to diverge; this phenomenon is known as diffraction. Given the wavelength of the light and the geometry of the apparatus (i.e., the separation and widths of the slits and the distance from the slits to the screen), one can use the wave theory to calculate the expected pattern in each case; the theory agrees precisely with the experimental data.

1.2 Basic Constants in Nature

There are three fundamental dimension constants in nature: the speed of light c, Planck's constant $\hbar = h/2\pi$, and Newton's constant gravity G_N. Their values are given by,

$$c = 2.99792 \times 10^8 \text{ m/s}$$
$$\hbar = 1.05457 \times 10^{-34} \text{ Js}$$
$$G_N = 6.6732 \times 10^{-11} \text{ Jm/(kg)}^2$$

Instead of Joules (J), one often uses electron-volts (eV) to express energy; the two are related by $1 \text{ eV} = 1.60219 \times 10^{-19}$ J. All other dimension constants are either artifacts of the units used (such as the Boltzmann constant k_B, which converts temperature to energy by $k_B T$) or composites (such as Stefan-Boltzmann's constant $\sigma = \pi^2 k^4 / 60 c^2 \hbar^3$).

Using c to convert a frequency ν into a wavelength λ by $\lambda = c/\nu$ and vice-versa, and $\hbar$ to convert frequency into energy E by $E = h\nu$ and vice-versa, all length scales, times, and energies may be converted into one another. Finally, with the help of Einstein's relation E = mc2, we may also convert a rest mass m into an energy E. It is, therefore, a common practice to describe masses, times, and length scales all in eV.

The role of Newton's gravitational constant GN is actually to set an absolute scale (the so-called Planck scale) of energy EP, (and thus of mass, length, and time), whose value is $E_P = (\hbar c/G_N)^{1/2} = 1.22 \times 10^{19}$ GeV. This energy scale is much larger than the highest energies reachable by particle accelerators today (only about 10^3 GeV). It corresponds to the energy where quantum effects in gravity become strong. In this course, gravitational effects will always be neglected.

CHAPTER 2:

The Principle of Wave-Particle Duality — An Overview

Beginning with the popular problem of blackbody radiation, physics entered a time of deep crisis in the year 1900 when a series of strange phenomena for which no traditional explanation was conceivable started to grow one after the other. By 1923, it was clear that all of these oddities had a common source. They discovered a novel fundamental theory of nature that contradicted classical physics: the well-known principle of wave-particle duality which can be expressed as follows.

The wave-particle duality theory is as follows. All physical objects have two personalities: waves and particles. Anything we used to think of as purely a wave, now has a corpuscular character, and anything we used to think of as purely a particle now behaves like a wave. These two traditionally irresolvable points of view, particle vs. wave, have a relationship,

$$E = hf, \quad p = h/\lambda \quad (1.1)$$

or, equivalently.

$$f = E/h, \quad \lambda = h/p \quad (1.2)$$

In expressions (1.1), we start off with what we traditionally considered to be solely a wave — an electromagnetic (EM) wave, for example — and we associate its wave characteristics f and λ (frequency and wavelength) with the corpuscular characteristics E and p (energy and momentum) of the corresponding particle. Conversely, in expressions (1.2), we begin with what we once regarded as purely a particle — say, an electron — and we associate its corpuscular characteristics E and p with the wave characteristics f and λ of the corresponding wave. Planck's constant h, which provides the link between these two aspects of all physical entities, is equal to:

$$h = 6.62 \times 10^{-27} \text{ erg s} = 6.62 \times 10^{-34} \text{ Js}$$

Actually, the aim here is not to retrace the historical process that led to this fundamental discovery, but precisely the opposite. Taking wave-particle duality for granted, we aim to show how effortlessly the peculiar phenomena we mentioned earlier can be explained. Incidentally, these phenomena merit discussion not only for their historical role in the discovery of a new physical principle but also because of their continuing significance as fundamental quantum effects. Furthermore, we show that the principle of wave-particle duality should be recognized as the only sensible explanation to fundamental "mysteries" of the atomic world, such as the extraordinary stability of its structures (e.g., atoms and molecules) and the uniqueness of their form, and not as some whim of nature, which we are supposed to accept merely as an empirical fact.

From its very name, it is clear that the principle of wave-particle duality can be naturally split into two partial principles:

(i) The Principle Of The Wave-Particle Duality Of Light And

(ii) The Principle Of The Wave-Particle Duality Of Matter

We proceed to examine both these principles concerning the peculiar phenomena and problems that led to their postulation.

2.1 The Principle of Wave-Particle Duality of Light

According to the preceding discussion, the wave-particle duality says that light — which in classical physics is purely an EM wave — also has a corpuscular character. The associated particle is the celebrated quantum of light, the photon. The wave-like features f and λ of the corresponding EM wave, and the particle-like features E and p of the associated particle, the photon, are related through expressions (1.1). We will now see how this principle can explain two key physical phenomena — the photoelectric effect and the Compton effect — that are completely inexplicable in the context of classical physics.

2.2 The Photoelectric Effect

With this term, we refer today to the general effect of light-induced removal of electrons from physical systems where they are bound. Such systems can be atoms and molecules, in which case we call the effect ionization or a metal; for that purpose, we have the standard photoelectric effect studied at the end of the nineteenth and the beginning of the twentieth century. What makes the effect peculiar from a classical perspective is the failure of classical physics to explain the following empirical fact: The photoelectric effect (i.e., the removal of electrons) is possible only if the frequency f of the incident EM radiation is greater than (or at least equal to) a value f_o that depends on the system from which the removal occurs (atom, molecule, metal, etc.).

We thus have:

$$f \geq f_o \qquad (1.3)$$

In classical physics, a "threshold condition" of the type (1.3) has no physical justification. Whatever the frequency of the incident EM wave, its electric field will always produce work on the electrons. When this exceeds the work function of the metal — the minimum energy required for extraction — electrons will be ejected from it. In other words, in classical physics, frequency plays no crucial role in the energy exchanges between light and matter while the intensity of the electric field of light is the decisive factor. Clearly, the very existence of a threshold frequency in the photoelectric effect leaves no room for a classical explanation.

In contrast, the phenomenon is easily understood in quantum mechanics. A light beam of frequency f is also a stream of photons with energy $\epsilon = hf$; therefore, when quantized light —

a "rain of light quanta" — impinges on metal, only one of two things can happen: Since the light quantum is by definition indivisible, when it "encounters" an electron it will either be absorbed by it or "pass by" without interacting with it. In the first case (absorption), the outcome depends on the relative size of $\epsilon = hf$ and the work function, W, of the metal. If the energy of the light quantum (i.e., the photon) is greater than the work function, the photoelectric effect occurs; if it is lower, there is no such effect. Therefore, the quantum nature of light points naturally to the existence of a threshold frequency in the photoelectric effect, based on the condition,

$$hf \geq W \Rightarrow f \geq W/h = f_0 \quad (1.4)$$

which also determines the value of the threshold frequency $f_0 = W/h$. For $hf > W$, the energy of the absorbed photon is only partially spent to extract the electron, while the remainder is turned into kinetic energy K ($= mv^2/2$) of the electron. We thus have

$$hf = W + K = W + 1/2\, mv^2 \quad (1.5)$$

which is known as Einstein's photoelectric equation. Written in the form.

$$K = hf - W \quad (f \geq f_0) \quad (1.6)$$

Equation (1.5) predicts a linear dependence of the photoelectrons' kinetic energy on the light frequency f, represented by the straight line in Figure 1.1. Therefore, by measuring K for various values of f, we can fully confirm, or disprove, Einstein's photoelectric equation and, concomitantly, the quantum nature of light, as manifested via the photoelectric effect. In addition, we can deduce the value of Planck's constant from the slope of the experimental line.

The discussion becomes clearer if in the basic relation $\epsilon = hf = hc / \lambda$ we express energy in electron volts and length in angstroms—the "practical units" of the atomic world (1Å = 10^{-10} m, 1 eV = 1.6×10^{-19} J = 1.6×10^{-12} erg). The product hc, which has dimensions of energy times length (since h has dimensions of energy times time), then takes the value hc = 12400 eVÅ, and the formula for the energy of the photon is written as:

$$\epsilon(eV) = 12400 / \lambda(Å) \approx 12000 / \lambda(Å) \quad (1.7)$$

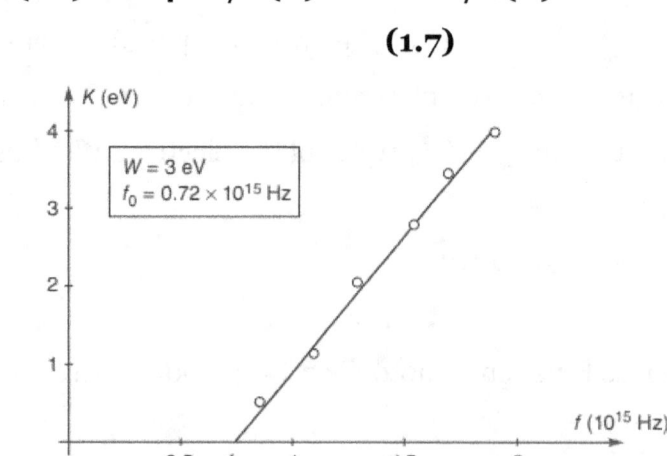

Figure 1.1 The kinetic energy K of electrons as a function of photon frequency f. The experimental curve is a straight line whose slope is equal to Planck's constant.

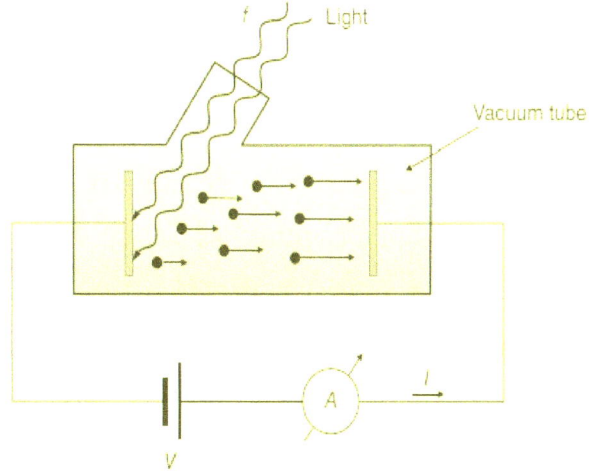

Figure 1.2 The standard experimental setup for studying the photoelectric effect. The photoelectric current occurs only when $f > f_0$ and vanishes when f gets smaller than the threshold frequency f_0. The kinetic energy of the extracted electrons is measured by reversing the polarity of the source up to a value V_0—known as the *cutoff potential*—for which the photoelectric current vanishes and we get $K = eV_0$.

The last expression is often used in this book since it gives simple numerical results for typical wavelength values. For example, for a photon with $\lambda = 6000$ Å — at about the middle of the visible spectrum — we have $\epsilon = 2$ eV. We remind the readers that the electron volt (eV) is defined as the kinetic energy attained by an electron when it is accelerated by a potential difference of 1 V. Figure 1.2 shows a typical setup for the experimental study of the photoelectric effect. Indeed, Einstein's photoelectric equation is validated by experiment, thus confirming directly that light is quantized as predicted by the principle of the wave-particle duality of light.

2.3 The Compton Effect

According to expressions (1.1), a photon carries energy $\epsilon = hf$ and momentum $p = h/\lambda$. And because it carries momentum, the photon can be regarded as a particle in the full sense of the term. But, how can we verify that a photon has not only energy but also momentum? Clearly, we need an experiment whereby photons collide with very light particles — we will shortly see why. We can then apply the conservation laws of energy and momentum during the collision to check whether photons satisfy a relation of the type $p = h/\lambda$.

Why do we need the target particles to be as light as possible — that is, electrons? It is well known that when small moving spheres collide with considerably larger stationary ones, they simply recoil with no significant change in their energy. In contrast, the large spheres stay practically still during the collision. Conversely, if the target spheres are small (or even smaller than the projectile particles), upon collision, they will move, taking some of the impinging spheres' kinetic energy, which then scatters in various directions with lower kinetic energy. Therefore, if photons are particles in the full sense of the term, they will behave as such when scattered by light particles, like the electrons of material. They will transfer part of their momentum and energy to the target electrons and end up with lower energy than they had before the collision. In other words, we will have:

$$\epsilon' = hf' < \epsilon = hf \Rightarrow f' < f \Rightarrow \lambda' > \lambda$$
(1.8)

This is where the primes refer to the scattered photons. This shift of the wavelength to greater values when photons collide with electrons is known as the Compton effect. It was confirmed experimentally by Arthur H. Compton in 1923 when an x-ray beam was scattered off by the electrons of a target material. Why were x-rays used to study the effect? (Today, we actually prefer γ rays for this purpose.) Because x- (and γ) rays have a very short wavelength, the momentum $p = h/\lambda$ of the impinging photons is large enough to ensure large momentum and energy transfer to the practically stationary target electrons (whereby the scattered photons suffer a great loss of momentum and energy). In a Compton experiment, we measure the wavelength λ' of the scattered photon as a function of the scattering angle θ between the directions of the impinging and scattered photon. By applying the principles of energy and momentum conservation, we can calculate the dependence $\lambda' = \lambda'(\theta)$ in a typical collision event such as the one depicted in Figure 1.3.

Indeed, if we use the conservation equations to eliminate the parameters E, p, and ϕ (which we do not observe in the experiment, as they pertain to the electron), we eventually obtain:

$$\Delta\lambda = \lambda' - \lambda = h/mc\,(1 - \cos\theta) = \lambda_C\,(1 - \cos\theta)$$
(1.9)

Where,

$$\lambda_C = h/mc = 0.02427 \, \text{Å} \approx 24 \times 10^{-3} \, \text{Å}$$
(1.10)

is the so-called Compton wavelength of the electron. It follows from (1.9) that the fractional shift in the wavelength, $\Delta\lambda/\lambda$, is on the order of λ_C/λ, so it is considerable in size only when λ is comparable to or smaller than the Compton wavelength. This condition is met in part for hard x-rays and in full for γ rays. Compton's experiment fully confirmed the prediction (1.9) and, concomitantly, the relation $p = h/\lambda$ on which it was based. The wave-particle duality of light is thus an indisputable experimental fact. Light — and, more generally, EM radiation — has a wave-like and a corpuscular nature at the same time.

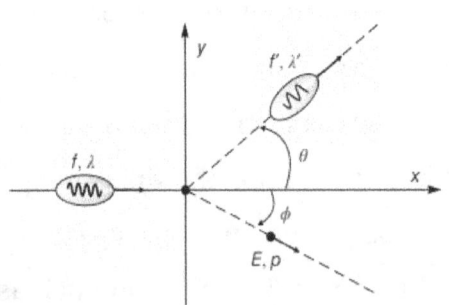

Figure 1.3 A photon colliding with a stationary electron. The photon is scattered at an angle θ with a wavelength λ' that is greater than its initial wavelength λ. The electron recoils at an angle ϕ with energy E and momentum p.

2.4 The Principle of Wave-Particle Duality of Matter

As emphasized in the introduction, relations (1.2), of the wave-particle duality of matter are similar to those of light — relations (1.1) — but they have to be viewed in reverse order. In case (1.2), we are talking about entities (e.g., electrons) we used to recognize as particles in classical physics (so they are described by their energy E and momentum p), but we now learn they are also waves. Their wave features f and λ are connected to the corpuscular attributes E and p via relations (1.2). The electron — the most fundamental particle of nonnuclear matter — is thus a particle and a wave simultaneously. We were already aware of its corpuscular nature; after all, we first came across the electron as a particle. So, we just need to examine if it is also a wave with $\lambda = hp$ as Louis de Broglie first hypothesized in 1923. Let us examine how we can infer the existence of these waves.

2.5 Energy Quantization in Matter Waves

The most important general consequence of wave-particle duality of the matter is to experimentally verify electrons' wave-like nature. The obvious test is to look for interference phenomena between electronic waves, just as in classical waves. This would be a direct confirmation. But there is also an indirect confirmation, invoking a characteristic feature of standing waves, namely, frequency quantization. A standing classical wave — localized on a finite object — can only exist if its frequency takes a discrete sequence of values known as the eigenfrequencies of the system. The most representative examples are the standing waves of definite frequency — the so-called normal modes — on a string. As it follows from Figure

1.4, the allowed frequencies of the string's vibrations—$f = c\lambda$, where c is the speed of wave propagation, is given by

$$L = n\lambda/2 \Rightarrow \lambda = 2L/n \Rightarrow f = c/\lambda = c/(2L/n) = c/2L \cdot n$$
(1.12)

which means that the only possible vibrations of the string are those with integer multiples of the fundamental frequency $f_1 = c/2L$.

But if the frequency is quantized in classical systems, too will be the particle's energy since the wave-particle duality of particles — namely, the relation $E = hf$ — provides a direct link between their energy the frequency of the corresponding wave. So, if a quantum particle, say an electron, is trapped somewhere in space (e.g., in an atom or a molecule), the associated de Broglie wave will be a standing wave with quantized frequency. Therefore, the energy $E = hf$ of the electron will also be quantized. As we will see shortly, energy quantization for particles trapped in some space region (and thus perform confined motion) is the deepest consequence of the wave-particle duality of matter.

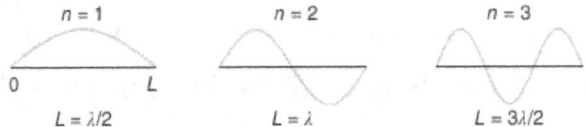

Figure 1.4 Standing classical waves on a string. A standing wave of this kind can only be formed when an integer number of half-waves fit on the string. That is, $L = n\frac{\lambda}{2}$ ($n = 1, 2, ...$).

2.6 Atomic Stability under Collisions

We can now see that the only logical scenario for the mystery of atomic stability is the energy quantization of atomic electrons. Why is this a puzzle Since atoms are totally unbothered even though they collide violently with one another on a regular basis?

Suppose we picture the electrons in atoms orbiting around the nucleus, like planets around a sun. In that case, it is as if their orbits do not change at all upon innumerable collisions with other "solar systems." But there is more to it. Even if we took apart an atom — by removing all its electrons — and let it "reconstruct" itself, it would reemerge in identical form and shape. The evidence for these statements is that atoms always emit the same characteristic frequencies — the same spectrum — while their chemical behavior remains unaltered. In fact, chemical stability is an essential prerequisite for our very existence. Note, however, that in the discussion, we only considered atomic stability against collisions. We have completely ignored the stability of atoms against the radiation emitted by their electrons, which, being charged particles in accelerated motion, ought to radiate and lose energy until they fall into the nucleus. Let us simply accept that, for some reason, the classical laws of EM radiation do not hold in the atomic world. Let us then see how atomic stability against collisions can be explained naturally by assuming that the energy of atomic electrons is quantized. In the hydrogen atom, for example, if the electron, which has quantized energy, occupies the lowest possible state — the so-called ground state — then the smallest possible change for the atom is a transfer of the electron from the ground state to the next available state, namely, the first excited state. In other words, the electron can only make a discontinuous transition — a quantum jump (or leap) — toward an excited atomic state. Now, if the environment offers less energy to the atom than the energy required for such a leap, as is the case for thermal collisions at room temperature, then no transition can occur at all. Indeed, the energy difference between the ground and first excited states of an atom — or molecule — is on the order of a few eV.

In contrast, the average thermal energy at room temperature is about 100th of this value. As a result, thermal collisions at room temperature do not provide sufficient energy to excite the atoms, which thus behave as stable and invariant entities during collisions. We have to reject the classically allowed small, gradual changes in energy and consider only those quantum jumps for which the minimum required energy is available. Hence, energy quantization arises as to the only conceivable explanation of the mystery of atomic stability. The "equation:"

Quantization = stability

emerges thus as the fundamental conceptual equation of quantum physics. And since the only natural mechanism of quantization, that we are aware of involves standing waves, the following reasoning also applies:

Stability → quantization → wave-like behavior

This explanation of the central mystery of the atomic world — the remarkable stability of its structures — demonstrates that the notion of wave-like behavior for particles is not so "crazy" after all. In hindsight, we can regard it as the only natural explanation of the most fundamental problem put forward by studying matter at the atomic level.

2.7 Energy Scales

The main idea of the previous discussion — namely, microscopic particles in confined motion inside a structure (such as an atom or a molecule) are represented by standing matter waves — helps us understand another central mystery of the atomic world: The smaller the region inside which a particle resides, the greater the energy of that particle. The most typical examples of this mystery are the atom and the nucleus. Atomic electrons (of the outer shell, for heavy atoms) have energies on a few eV orders. In contrast, the corresponding energies for protons and neutrons inside the nucleus are one million times greater — that is, on the order of a few MeV! Again, the explanation lies in the wave-particle duality expression $\lambda = hp$ and the realization that the first (fundamental) standing wave in a region of space — recall the example of the string — has a wavelength λ on the order of the linear size of the region. The wavelengths of the higher standing waves are even smaller. So, we can say that the largest wavelength — the one that corresponds to the ground state — will be about the size of:

$$\lambda_{max} \approx 2L$$

where L is the linear size of the region within which the standing wave is formed. In this case, the relation $\lambda = hp \Rightarrow p = h/\lambda$ shows that the momentum of the trapped wave-particle cannot be smaller than

$$p_{min} = h/\lambda_{max} \approx h/2L$$

And if we are interested in the state of lowest energy — which is certainly the most important state — then $p \approx p_{min}$, and the formula:

$$p \approx h/2L$$

provides a good estimate of the momentum of particles trapped inside a quantum system of linear dimension L. For the corresponding kinetic energy ($p^2/2m$) of these particles, we have:

$$K \approx h^2/8mL^2 \qquad (1.13)$$

The conclusion is now clear: The smaller the region inside which a particle is moving, the smaller the wavelength (in the first standing wave) of that particle is and, consequently, the greater its momentum and energy. Figure 1.5 should help visualize the physics of this key fact.

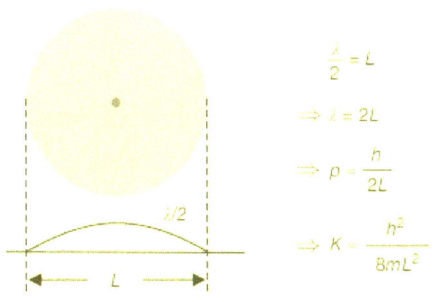

Figure 1.5 A standing matter wave of spherical shape. A particle trapped inside a bounded region—a spherical volume in our case—of linear dimension L, is described (in the state of lowest energy) by a spherical standing wave that vanishes only at the boundary of this region. For its wavelength we thus have $\lambda/2 = L \Rightarrow \lambda = 2L$.

If we now apply the formula (1.13) on a nucleon — where $m = m_N$ =mass of a proton or neutron, and $L \approx 2R$ (R =nuclear radius) — or an electron of an outer atomic shell — where m = me and L = 2a (a =atomic radius) — we obtain:

$$K_N = h^2 / 32m_N R^2 = h^2 / 32m_e a^2 \cdot a^2 / R^2 \cdot m_e / m_N = K_e (a/R)^2 m_e / m_N \quad (1.14)$$

Given now that a ≈ 1 Å ≈ 10^{-10} m, $R \approx 10^{-15}$ m, and $m_N \approx 1836\, m_e$, expression (1.14) yields:

$$K_N \approx (10^6 - 10^7)\, K_e \quad (1.15)$$

which tells us that the kinetic energies K_N of protons and neutrons inside the nucleus are a few million times greater than the kinetic energies K_e of the outer-shell electrons in atoms. (Inner-shell electrons have greater energies than electrons in the outer shells since they move in a smaller region of space.) Furthermore, we can use the formula $K_e = h^2/32m_e a^2$ to obtain a typical value for the kinetic energy of the outer electrons:

$$K_e \approx \text{a few eV} \quad (1.16)$$

We combine Eq. (1.15) with Eq. (1.16) to obtain:

$K_N \approx$ a few MeV

$$(1.17)$$

If we now take the next logical step, namely, the energies released in chemical and nuclear reactions should be on the same order of magnitude as the energies of outer-shell atomic electrons and nucleons, respectively, then we can deduce another fundamental feature of our world: Energies released in chemical reactions can only be on the order of an eV. In contrast, energies released in nuclear reactions must be on the order of a MeV per reaction. We can thus say that eV and MeV define the chemical and nuclear energy scales, respectively.

We can now reexamine the problem of atomic stability. Suppose the energy scale of electrons in atoms — in the hydrogen atom, for simplicity — is on an eV order. In that case, differences between adjacent energy levels (remember, they are quantized) should be of the same order; that is, a few eV. Note, for example, that the atom's first excited state will correspond to a standing matter wave with $\lambda = L$ (one-half that of the ground state), so the electronic momentum $p = h/\lambda$ will double, and the kinetic energy will quadruple compared to the ground state. (Provided, of course, that all standing waves of the atom occur within the same volume in space, which is not exactly true; in its excited states, the atom is bigger.) The energy difference between the ground and first excited states of an atom, such as hydrogen, would thus also be on the order of an eV; this energy difference determines the atom's stability against collisions, as we noted earlier. We remind the readers that the factor kT determines the average magnitude of thermal energies at temperature T via the relation:

$$K = 3/2\, kT \tag{1.18}$$

takes (at room temperature) the approximate value.

$$kT|_{T\approx 300\,K} \approx 1/40 \text{ eV} \tag{1.19}$$

Thus, we can see that thermal collisions at room temperature and at much higher temperatures, say a few thousand degrees, cannot cause atomic excitations. Atoms emerge from their incessant collisions — roughly one billion collisions per second, as we saw — completely intact. In reality, not all atoms of gas have the same thermal kinetic energy — (1.18) is merely a mean value — but obey a Maxwell–Boltzmann distribution, so some are much more energetic than others and able to cause mutual excitations when they collide with each other. So, the exact picture is this: Even at room temperature, a small fraction of atoms in a gas are excited, but the overwhelming majority remains intact in their ground state.

In the case of an atomic nucleus, where the energy difference between the ground and first excited states is on a MeV order, similar reasoning leads us to conclude that nuclear stability against collisions is a million times greater than atomic stability. Thus, the critical temperature for a nucleus's stability is a few billion degrees kelvin compared to a few thousand degrees for an atom. Therefore, for thermonuclear reactions to occur, as in a star's interior, the temperature needs to rise to billions of degrees! And yet thermonuclear reactions inside stars occur — for without such reactions, we would not exist — even though the typical temperature in their interior is no greater than 10–20 million degrees!

2.8 The Stability of Atoms and Molecules Against External Electromagnetic Radiation

There are two types of external "perturbations" that atoms and molecules are often subjected to and which could threaten their structural stability. The first perturbation is thermal collisions — actually, electric forces between electrons of approaching atoms — which we have already examined. The second type of perturbation is the ubiquitous electromagnetic radiation — visible light, infrared (IR), UV, x-rays, radio waves, and so on — that hits atoms continuously. Does EM radiation change the structure of atoms? If the atoms were classical systems, then the answer would surely be in the affirmative since they would have to "respond" to any external perturbation, however small, by changing their structure accordingly; for example, by slightly adjusting their electronic orbits. However, atoms are not classical but quantum systems, and therefore their states have quantized energies that can only change via specific quantum jumps. In other words, atoms cannot absorb an arbitrary amount of energy but only the amount required for a transition from the ground state — if this is where they start from — to any one of their excited states. Now, due to the wave-particle duality of light, the incident EM radiation on an atom is also quantized with an energy quantum equal to:

$$\epsilon = hf = hc / \lambda \Rightarrow (eV) \approx 12\,000 / \lambda(\text{Å}) \tag{1.20}$$

For example, for visible light, where

$$4000\,\text{Å} < \lambda < 7400\,\text{Å} \quad \text{(visible light)} \tag{1.21}$$

the energies of optical photons span the range:

$$1.6\,\text{eV} < \epsilon < 3\,\text{eV} \quad \text{(visible light)} \tag{1.22}$$

with a typical value — for $\lambda \approx 6000$ Å — equal to 2 eV. Thus, UV photons — being more energetic and more chemically potent than optical photons — have energies greater than 3 eV, while IR photons have energies less than 1.6 eV. In other words, UV- and IR-light photon energies lie to the right and left, respectively, of the visible range (1.22). For radio waves — where $f \approx 100\,\text{MHz} \Rightarrow \lambda = c/f = 3\,\text{m} = 3 \times 10^{10}\,\text{Å}$ —we have

ϵ **(radio waves) = 12000 / 3 × 10^{10} eV = 0.4 × 10^{-6} eV ≈ 1 μeV**

What happens when one of the kinds mentioned above of radiation impinges on an atom? Take, for example, the hydrogen atom, for which the first excitation energy — equal to the energy difference between its first excited and ground states — is 10.2 eV. Clearly, any radiation whose photons have energies less than 10.2 eV cannot induce any changes to the hydrogen atom. The photons of the impinging radiation "bounce off" the atom in another direction; they are scattered, as we say, leaving the atom intact. Hence, we can conclude that energy quantization of atomic electrons — and the corresponding energy scale on the order of an eV — combined with light quantization, ensures atomic stability against not only collisions but also all of EM radiation with energy below the UV: visible, IR, microwaves, radio waves, and so on. No matter how long atoms or molecules are bombarded by such radiation — provided its intensity is not too high — they remain completely unaffected. Similarly, radiation at such frequencies cannot cause chemical reactions. The reason is that, for chemical reactions, there is also an energy threshold, a minimum energy barrier the light quantum has to surpass for a reaction to occur. And just like the typical energies in atoms and (small) molecules are on the order of a few eV, this threshold energy is also on the order of a few eV, typically greater than 3 eV. So, the only kinds of radiation that are chemically potent are those in the UV range and beyond (x-rays, γ rays, etc.). This means, among other things, that visible light is not chemically dangerous, for if it were, we would not be here (since our planet is awash with it)!

We thus come to the realization that the crucial feature of the photoelectric effect — the existence of a threshold frequency (and energy) for the phenomenon to occur — is completely general. It holds for chemical reactions, excitations, dissociations of molecules, and so on. Consequently, all radiation with photon energies below the energy threshold is "harmless" in the sense that it cannot cause the abovementioned effects. Given also that all "threshold" energies are on the order of a few eV; atoms and molecules are completely "safe" against all radiation from the visible range and below (in energy). (Visible light can actually cause some reactions, but these belong to a very specific category.)

2.9 Probabilistic Interpretation of Matter Waves

In 1911, the Rutherford experiment showed that the atom consists of a tiny nucleus with electrons orbiting far away, like planets around the sun. No classical model could explain how such an atom may last more than a few tenths of a nanosecond! Whatever the electrons' classical orbit, their motion would surely be an accelerated one (with linear or centripetal acceleration). As a result, electrons would emit EM radiation continuously, lose energy, and ultimately fall — in an infinitesimally small amount of time — on the nucleus. Conclusion: A truly classical atom cannot exist. But can the quantum atom — based on the principle of the wave-particle duality of electrons — also solve the mystery of the atoms' stability against their own radiation, as it solved the two previous mysteries (stability against collisions and stability against external radiation)? Here, the answer is not a resounding yes as it was for the other two questions. At this point, the mystery of the stability of atoms against their own radiation cannot be solved directly because the quantum theory has not been "set up" yet. Nevertheless, this problem can be bypassed with the following reasoning: If the principle of wave-particle duality is correct, then the orbital motion of electrons (which is where radiation comes from) has no physical meaning. Let us elaborate: Orbital motion means that the electron at any given time is found at a specific location (i.e., localized in space), whereas the very concept of a wave postulates a physical entity that is spread out in space. Moreover, a particle is by definition "indivisible" — within certain limits — while a wave can always be divided, for example, by letting a part of it be transmitted through one slit and another part through another. A wave is thus always extended and divisible; a particle is always localized and indivisible. At this point, the reader would be justified to ask: How can the principle of the wave-particle duality of the matter then be true? How can we say that a particle is at the same time a wave? How can we fit within the same physical entity two mutually exclusive properties, such as "localized and indivisible" on the one hand and "extended and divisible" on the other? We have just arrived at the most critical question of quantum theory — a question that leads to the celebrated statistical (or probabilistic) interpretation of matter waves. Here is what this interpretation says (Max Born, 1927):

The function $\psi = \psi(\boldsymbol{r})$ that describes a matter wave (its so-called wave function) does not represent a measurable physical quantity. It is rather a mathematical wave — a probability wave — whose squared amplitude $|\psi(\boldsymbol{r})|^2$ yields the probability per unit volume to locate the particle in the vicinity of an arbitrary point r. We thus have:

$$P(r) = |\psi(r)|^2 \qquad (1.26)$$

$P(r)$ is the probability per unit volume — the probability density — of locating the particle in the vicinity of an arbitrary point in space. The total probability of finding the particle anywhere in space is given by the overall integral space:

$$\int |\psi(r)|^2 \, dV = 1 \qquad (1.27)$$

which clearly equals unity.

Given this interpretation, the wave function ψ has no immediate physical meaning — as it does not represent some sort of a physical wave, so it can take complex values in general. This is why absolute values are necessary for (1.26) or (1.27). According to (1.26), quantum particles frequent locations where their wave is strong — "stormy" areas — and avoid "calm" places where their wave is weak. In the context of such an interpretation, the contradiction between particles and waves is removed at once since the particle need not cease being a particle and does not have to physically "disperse" throughout the volume of the wave. The wave simply describes the probability of detecting the particle here or there, but never here and there at the same time. When we do locate the particle, our detectors always record an integral and indivisible entity. No experiment has ever "captured" half an electron or a quarter of a proton. To give readers an idea of how we describe quantum particles, we depict in Figure 1.7 two simple examples of one-dimensional wave functions. This short detour in our discussion helped us arrive at the following basic conclusion: The correct interpretation of the principle of wave-particle duality strips the concept of electronic orbits in atoms of any physical meaning. As a result, it makes no sense to speak of accelerated motion of electrons, nor, therefore, of emission of radiation from them. In other words, we do not have a solution to the problem mentioned earlier — but we do not have a problem either!

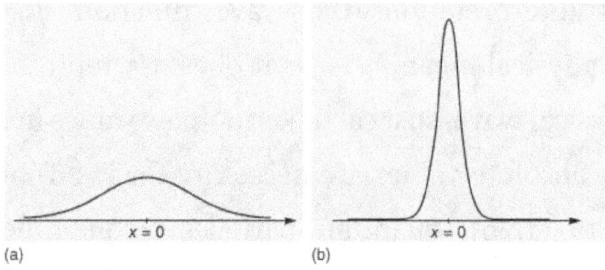

Figure 1.7 Typical one-dimensional wavefunctions. (a) An extended wavefunction: The position of the particle is known with very low precision. There is a significant probability of locating the particle in regions away from the "most frequented" location at $x = 0$.
(b) A localized wavefunction: The position of the particle is known with very high precision. In the vast majority of the measurements, we would detect the particle in the immediate vicinity of $x = 0$.

2.10 Quantum Jumps from Higher to the Lower Energy States and Atomic Spectra

Let us make some "impromptu" thoughts on this question using the hydrogen atom again as an example. Like any standing wave, a standing electron wave around the nucleus can exist in a number of possible forms — the so-called normal modes. The first form corresponds to the lowest energy state, and the next one corresponds to excited atomic states. The corresponding energies are quantized according to some discrete sequence E_1, E_2,..., E_n,... Since these successive standing waves around the nucleus represent the only possible energy states of the electron, we have the following two scenarios:

a) If the electron is in the ground state, it obviously cannot radiate; for if it did, it would lose energy and move to a lower energy state, which, however, does not exist.

b) If the electron is in an excited state — say, the first excited state — it can be de-excited, but only according to the basic quantum rules described earlier. First of all, a gradual de-excitation is impossible because the electron would then be able to gradually shed its excess energy in the form of radiation and transit to states with gradually decreasing energy, which, again, do not exist. The only available state to go to is the ground state, which is located (in the hydrogen atom) 10.2 eV below the first excited state. So, what can the excited electron do to "shed" its excess energy and return to the ground state? Very simply, a quantum jump: an abrupt transition from the excited to its ground state via emission of the energy difference 10.2 eV in the form of a UV photon.

Atoms, therefore — and, likewise, molecules, and all other quantum systems — emit light only when they undergo a transition from one of their excited states to a lower state. When this happens, a photon is emitted with energy hf, equal to the energy difference between the initial and final states of the transition. We thus have:

$$E_n - E_m = hf_{nm} \quad (n > m) \tag{1.28}$$

where E_n ($n > 1$) is the energy of the initial excited state of the atom, and E_m is the energy of the final state (which may or may not be its ground state). The frequencies f_{nm} are what we observe in the so-called line emission or absorption spectrum of a gas made of the atoms or molecules we wish to study. We thus realize that the quantization of electronic energies in atoms or molecules is reflected in the line spectra of the corresponding substances in gas form. In turn, these spectra are our best "tool" for measuring the allowed energies in a quantum system. The frequencies f_{nm} that correspond to the transitions $n \to m$ are known as Bohr frequencies. Theoretical physicists, however, prefer to use the same term for the corresponding angular frequencies, $\omega_{nm} = 2\pi f_{nm}$, because ω is better suited than f ($= \omega/2\pi$) for the mathematical description of harmonic oscillations or waves. Note, for example, that the mathematical expression of a harmonic oscillation $x(t) = A \sin(2\pi t/T)$ — where T is the period — takes the much simpler form $x(t) = A \sin\omega t$ if we introduce the angular frequency ω, via the relation:

$$\omega = 2\pi / T = 2\pi f \tag{1.29}$$

In the same spirit, theoretical physicists prefer to write the fundamental relation $\epsilon = hf$ in the equivalent form:

$$\epsilon = hf = h\,\omega/2\pi = h\,2\pi/\omega = \hbar\omega,$$

Where,

$$\hbar = h / 2\pi \tag{1.30}$$

is the so-called reduced Planck's constant. As we will see later, the mathematical expressions of basic quantum results are considerably simplified when written in terms of $\hbar$ instead of h. Thus, the use of $\hbar$ instead of h is now common in quantum physics. At the same time, one can always revert to the older symbol whenever there is a need to use quantities closer to what is experimentally measured, such as the frequency f or the wavelength λ.

Having thus opted to use $\hbar$ over h, we can rewrite the second expression — $p = h/\lambda$ — of the wave-particle duality as:

$$\mathbf{p} = h/\lambda = 2\pi h/\lambda = h 2\pi/\lambda = \hbar \mathbf{k} \tag{1.31}$$

where,

$$\mathbf{k} = 2\pi/\lambda \tag{1.32}$$

is the so-called wavenumber of the wave. Clearly, k is the spatial equivalent of ω, with λ in place of T, as we should have expected since λ is the spatial and T is the temporal period of a sinusoidal wave. The modern version of the wave-particle duality is thus written as:

$$E = \hbar\omega, \quad \mathbf{p} = \hbar\mathbf{k}$$

which is clearly more elegant than the older form.

CHAPTER 3:

Concepts of Quantum Physics

The chapter provides in-depth information on the various concepts of Physics and lets you understand the researches and laws with complete information to clarify many laws of nature

3.1 Planck's Radiation Law

Physicists almost unanimously embraced the wave theory of light by the end of the nineteenth century. However, while classical physics theories describe interference and diffraction processes in light propagation, they do not account for light absorption and emission. All bodies emit electromagnetic energy as heat; in reality, all wavelengths of radiation are emitted by a body. The amount of energy radiated at various wavelengths reaches a peak at a wavelength that varies with the body's temperature; the hot body the short-wavelength shows the maximum radiation. Attempts to use classical ideas to quantify the energy distribution for radiation from a blackbody were unsuccessful. (A blackbody is a hypothetical ideal body or surface that absorbs and reemits all radiant energy falling on it.) One formula, proposed by Wilhelm Wien of Germany, did not agree with observations at long wavelengths. Another, proposed by Lord Rayleigh (John William Strutt) of England, disagreed with those at short wavelengths.

In 1900, the German theoretical physicist Max Planck made a bold suggestion. He assumed that the radiation energy is emitted, not continuously, but rather in discrete packets called quanta. The energy E of the quantum is related to the frequency ν by $E = h\nu$. The quantity h, now known as Planck's constant, is a universal constant with the approximate value of 6.62607×10^{-34} joule·second. Planck showed the calculated energy spectrum, then agreed with observation over the entire wavelength range.

3.2 Einstein and the Photoelectric Effect

In 1905, Einstein extended Planck's Hypothesis to explain the photoelectric effect, which is the emission of electrons by a metal surface irradiated by light or more energetic photons. The kinetic energy of the emitted electrons depends on the frequency ν of the radiation, not on its intensity; for a given metal, there is a threshold frequency ν0 below which no electrons are emitted. Furthermore, emission occurs as soon as the light shines on the surface; there is no detectable delay. Einstein showed that these results can be explained by two assumptions: (1) that light is composed of corpuscles or photons, the energy given by Planck's relationship, and (2) that an atom in the metal can absorb either a whole photon or nothing. Part of the energy of the absorbed photon frees an electron, which requires a fixed energy W, known as the work function of the metal; the rest is converted into the kinetic energy $m_e u^2/2$ of the emitted electron (m_e is the mass of the electron and u is its velocity). Thus, the energy relation is a special composition for the article "Quantum Mechanics." If v is less than v_0, where $hv_0 = W$, no electrons are emitted. Not all the experimental results mentioned above were known in 1905, but all Einstein's predictions have been verified.

3.3 Bohr's Theory of the Atom

Niels Bohr of Denmark made a major contribution to the subject, who applied the quantum hypothesis to atomic spectra in 1913. The spectra of light emitted by gaseous atoms had been studied extensively since the mid-19th century. It was found that radiation from gaseous atoms at low pressure consists of a set of discrete wavelengths. This is quite unlike the radiation from a solid, distributed over a continuous range of wavelengths. The set of discrete wavelengths from gaseous atoms is known as a line spectrum because the radiation (light) emitted consists of a series of sharp lines. The wavelengths of the lines are characteristic of the element and may form extremely complex patterns. The simplest spectra are atomic hydrogen and the alkali atoms (e.g., lithium, sodium, and potassium). For hydrogen, the wavelengths λ are given by the empirical formula special composition for the article "Quantum Mechanics," where m and n are positive integers with n > m and $R\infty$, known as the Rydberg constant, have the value 1.097373157 × 107 per meter. For a given value of m, the lines for varying n form a series. The lines for m = 1, the Lyman series, lie in the ultraviolet part of the spectrum; those for m = 2, the Balmer series, lie in the visible spectrum; and those for m = 3, the Paschen series, lie in the infrared.

Bohr started with a model suggested by the New Zealand-born British physicist Ernest Rutherford. The model was based on Hans Geiger and Ernest Marsden's experiments, who, in 1909, bombarded gold atoms with massive, fast-moving alpha particles; when some of these particles were deflected backward, Rutherford concluded that the atom has a massive, charged nucleus. In Rutherford's model, the atom resembles a miniature solar system with the nucleus acting as the Sun and the electrons as the circulating planets. Bohr made three assumptions. First, he postulated that in contrast to classical mechanics, where an infinite number of orbits is possible, an electron can only be one of a discrete set of orbits, which he termed "stationary states." Second, he postulated that the only orbits allowed are those for which the electron's angular momentum is a whole number of n times $\hbar$ ($\hbar$ = h/2π). Third, Bohr assumed that Newton's motion laws, so successful in calculating the paths of the planets around the Sun, also applied to electrons orbiting the nucleus. The force on the electron (the analog of the gravitational force between the Sun and a planet) is the electrostatic attraction between the positively charged nucleus and the negatively charged electron.

With these simple assumptions, he showed that the orbit's energy has formed the special composition for the article "Quantum Mechanics," where E_0 is a constant that may be expressed by a combination of the known constants e, me, and h. While in a stationary state, the atom does not give off energy as light; however, when an electron makes a transition from a state with energy En to one with lower energy Em, a quantum of energy is radiated with frequency ν given by the special equation composition for the article "Quantum Mechanics" Inserting the expression for En into this equation and using the relation λν = c, where c is the speed of light; Bohr derived the formula for the wavelengths of the lines in the hydrogen spectrum with the correct value of the Rydberg constant.

Bohr's theory was a brilliant step forward. Its two most important features have survived in present-day quantum mechanics. They are (1) the existence of stationary, non-radiating states and (2) the relationship of radiation frequency to the energy difference between the initial and final states in a transition. Before Bohr, physicists had thought that the radiation frequency would be the same as the electron's frequency of rotation in orbit.

3.4 Scattering of X-rays

Soon scientists were faced with the fact that another form of radiation, X-rays, also exhibits both wave and particle properties. Max von Laue of Germany had shown in 1912 that crystals can be used as three-dimensional diffraction gratings for X-rays; his technique constituted the fundamental evidence for X-rays' wave-like nature. The atoms of a crystal, which are arranged in a regular lattice, scatter the X-rays. For certain directions of scattering, all the crests of the X-rays coincide. (The scattered X-rays are said to be in phase and to give constructive interference.) For these directions, the scattered X-ray beam is very intense. Clearly, this phenomenon demonstrates the wave behavior. In fact, given the interatomic distances in the crystal and the directions of constructive interference, the wavelength of the waves can be calculated.

In 1922, the American physicist Arthur Holly Compton showed that X-rays scatter from electrons as if they were particles. Compton performed a series of experiments on the scattering of monochromatic, high-energy X-rays by the graphite. He found that part of the scattered radiation had the same wavelength λ_0 as the incident X-rays but with an additional component with a longer wavelength λ. To interpret his results, Compton regarded the X-ray photon as a particle that collides and bounces off an electron in the graphite target as though the photon and the electron were a pair of (dissimilar) billiard balls. Application of the laws of conservation of energy and momentum to the collision leads to a specific relation between the amount of energy transferred to the electron and the angle of scattering.

For X-rays scattered through an angle θ, the wavelengths λ and λ_0 are related by the special equation composition for the article "Quantum Mechanics." The experimental correctness of Compton's formula is direct evidence for the corpuscular behavior of radiation.

3.5 De Broglie's Wave Hypothesis

Faced with evidence that electromagnetic radiation has both particle and wave characteristics, Louis-Victor de Broglie of France suggested a great unifying hypothesis in 1924. De Broglie proposed that matter has wave as well as particle properties. He suggested that material particles behave like waves and that their wavelength λ is related to the linear momentum p of the particle by $\lambda = h/p$.

In 1927, Clinton Davisson and Lester Germer of the United States confirmed de Broglie's Hypothesis for electrons. Using a nickel crystal, they diffracted a beam of monoenergetic electrons. They showed that the waves' wavelength is related to the momentum of the electrons by the de Broglie equation. Since Davisson and Germer's investigation, similar experiments have been performed with atoms, molecules, neutrons, protons, and many other particles. All behave like waves with the same wavelength-momentum relationship.

3.6 Basic Concepts and Methods

Bohr's theory, which assumed that electrons moved in circular orbits, was extended by the German physicist Arnold Sommerfeld and others to include elliptic orbits and other refinements. Attempts were made to apply the theory to more complicated systems than the hydrogen atom. However, the ad hoc mixture of classical and quantum ideas made the theory and calculations increasingly unsatisfactory. Then, in the 12 months started in July 1925, a period of creativity without parallel in physics' history, there appeared a series of papers by German scientists that set the subject on a firm conceptual foundation. The papers took two approaches: (1) matrix mechanics, proposed by Werner Heisenberg, Max Born, Pascual Jordan, and (2) wave mechanics, put forward by Erwin Schrödinger. The protagonists were not always polite to each other. Heisenberg found the physical ideas of Schrödinger's theory "disgusting," and Schrödinger was "discouraged and repelled" by the lack of visualization in Heisenberg's method. However, Schrödinger, not allowing his emotions to interfere with his scientific endeavors, showed that the two theories are equivalent mathematically, despite apparent dissimilarities. The present discussion follows Schrödinger's wave mechanics because it is less abstract and easier to understand than Heisenberg's matrix mechanics.

3.7 Schrödinger's Wave Mechanics

Schrödinger expressed de Broglie's Hypothesis concerning the wave behavior of matter in a mathematical form that is adaptable to various physical problems without additional arbitrary assumptions. He was guided by a mathematical formulation of optics. The straight-line propagation of light rays can be derived from wave motion when the wavelength is small compared to the dimensions of the apparatus employed. Similarly, Schrödinger set out to find a wave equation for the matter that would give particle-like propagation when the wavelength becomes comparatively small. According to classical mechanics, if a particle of mass me is subjected to a force such that its potential energy is V (x, y, z) at position x, y, z, then the sum of V(x, y, z) and the kinetic energy p2/2me is equal to a constant, the total energy E of the particle. Thus, the special composition for the article "Quantum Mechanics."

It is assumed that the particle is bound — i.e., confined by the potential — to a certain region in space because its energy E is insufficient for it to escape. Since the potential varies with position, two other quantities also: the momentum and, hence, by extension from the de Broglie relation, the wavelength of the wave. Postulating a wave function Ψ (x, y, z) that varies with position, Schrödinger replaced p in the above energy equation with a differential operator that embodied the de Broglie relation. He then showed that Ψ satisfies the partial differential equation special composition for the article "Quantum Mechanics."

This is the (time-independent) Schrödinger wave equation, which established quantum mechanics in a widely applicable form. An important advantage of Schrödinger's theory is that no further arbitrary quantum conditions need to be postulated. The required quantum results follow from certain reasonable restrictions placed on the wave function. For example, it should not become infinitely large at large distances from the center of the potential.

Schrödinger applied his equation to the hydrogen atom, for which the potential function, given by classical electrostatics, is proportional to $-e2/r$, where $-e$ is the charge on the electron. The nucleus (a proton of charge e) is situated at the origin, and r is the distance from the origin to the electron's position. Schrödinger solved the equation for this particular potential with straightforward, though not elementary, mathematics. Only certain discrete values of E lead to acceptable functions Ψ. These functions are characterized by a trio of integers n, l, m, termed quantum numbers. E's values depend only on the integers n (1, 2, 3, etc.) and are identical with those given by the Bohr theory. The quantum numbers l and m are related to the electron's angular momentum; Square root of $\sqrt{l}\,(l + 1)\hbar$ is the magnitude of the angular momentum, and $m\hbar$ is its component along some physical direction.

The square of the wave function, $\Psi 2$, has a physical interpretation. Schrödinger originally supposed that the electron was spread out in space and that its density at point x, y, z was given by the value of $\Psi 2$ at that point. Almost immediately, Born proposed the accepted interpretation — namely, $\Psi 2$ gives the probability of finding the electron at x, y, z. The distinction between the two interpretations is important. If $\Psi 2$ is small at a particular position, the original interpretation implies that a small fraction of an electron will always be detected. In Born's interpretation, nothing will be detected there most of the time, but it will be a whole electron when something is observed. Thus, the electron's concept as a point particle moving in a well-defined path around the nucleus is replaced in wave mechanics by clouds that describe the probable locations of electrons in different states.

Schrodinger wave equation, or just Schrodinger equation, is one of the most fundamental equations of quantum physics, and an important topic for JEE. The equation, also called the Schrodinger equation, is basically a differential equation and is widely used in Chemistry and Physics to solve problems based on the atomic structure.

Schrodinger wave equation describes a particle's behavior in a field of force or the change of a physical quantity over time. Erwin Schrödinger, who developed the equation, was even awarded the Nobel Prize in 1933.

What is Schrodinger Wave Equation?

Schrodinger wave equation is a mathematical expression describing the electron's energy and position in space and time, considering the matter wave nature of the electron inside an atom.

It is based on three considerations. They are:

- Classical plane wave equation,
- Broglie's Hypothesis of matter-wave, and
- Conservation of Energy.

Schrodinger equation gives us a detailed account of the form of the wave functions or probability waves that control some smaller particles' motion. The equation also describes how these waves are influenced by external factors. Moreover, the equation uses the energy conservation concept that offers details about the behavior of an electron that is attached to the nucleus.

Besides, by calculating the Schrödinger equation, we obtain Ψ and Ψ2, which helps us determine the quantum numbers and the orientations and the shape of orbitals where electrons are found in a molecule or an atom.

There are two equations, which are a time-dependent Schrödinger equation and a time-independent Schrödinger equation.

Time-dependent Schrödinger equation is represented as:

$$i\hbar \frac{d}{dt}|\Psi(t)\rangle = \hat{H}|\Psi(t)\rangle$$

Time-dependent Schrödinger equation in position basis is given as:

$$i\hbar \frac{\partial \Psi}{\partial t} = -\frac{\hbar^2}{2m}\frac{\partial^2 \Psi}{\partial x^2} + V(x)\Psi(x,t) = \tilde{H}\Psi(x,t)$$

Where,

i = imaginary unit, Ψ = time-dependent wave function, h2 is h-bar, V(x) = potential and H^ = Hamiltonian operator.

3.8 Time-Dependent Schrödinger Equation

At the same time that Schrödinger proposed his time-independent equation to describe the stationary states, he also proposed a time-dependent equation to describe how a system changes from one state to another. By replacing the energy E in Schrödinger's equation with a time-derivative operator, he generalized his wave equation to determine the time variation of the wave function and its spatial variation. The time-dependent Schrödinger equation reads special composition for the article "Quantum Mechanics:" Schrodinger equation. The quantity i is the square root of −1. The function Ψ varies with time t as well as with position x, y, z. For a system with constant energy, E, Ψ has the special form composition for the article "Quantum Mechanics," where exp stands for the exponential function. The time-dependent Schrödinger equation reduces to the time-independent form.

The probability of a transition between one stationary atomic state and some other state can be calculated with the time-dependent Schrödinger equation's aid. For example, an atom may change spontaneously from one state to another with less energy, emitting the difference in energy as a photon with a frequency given by the Bohr relation. If electromagnetic radiation is applied to atoms and if the frequency of the radiation matches the energy difference between two stationary states, transitions can be stimulated. In a stimulated transition, the atom's energy may increase — i.e., the atom may absorb a photon from the radiation — or the energy of the atom may decrease, with the emission of a photon, which adds to the energy of the radiation. Such stimulated emission processes form the basic mechanism for the operation of lasers. The probability of a transition from one state to another depends on the l, m, ms quantum numbers of the initial and final states. For most values, the transition probability is effectively zero. However, for certain changes in the quantum numbers, summarized as selection rules, there is a finite probability. For example, according to one important selection rule, the l value changes by unity because photons have a spin of 1. The selection rules for radiation relate to the angular momentum properties of the stationary states. The absorbed or emitted photon has its own angular momentum, and the selection rules reflect the conservation of angular momentum between the atoms and the radiation.

3.9 Electron Spin and Antiparticles

In 1928, the English physicist Paul A.M. Dirac produced a wave equation for the electron that combined relativity with quantum mechanics. Schrödinger's wave equation does not satisfy the special theory of relativity requirements because it is based on a nonrelativistic expression for kinetic energy ($p^2/2m_e$). Dirac showed that an electron has an additional quantum number ms. Unlike the first three quantum numbers, ms is not a whole integer and can have only the values +1/2 and −1/2. It corresponds to an additional form of angular momentum ascribed to a spinning motion. (The angular momentum mentioned above is due to the orbital motion of the electron, not its spin.) The concept of spin angular momentum was introduced in 1925 by Samuel A. Goudsmit and George E. Uhlenbeck, two graduate students at the University of Leiden, Neth., to explain the magnetic moment measurements made by Otto Stern and Walther Gerlach of Germany several years earlier. The magnetic moment of a particle is closely related to its angular momentum; if the angular momentum is zero, so is the magnetic moment. Yet, Stern and Gerlach had observed a magnetic moment for electrons in silver atoms, which were known to have zero orbital angular momentum. Goudsmit and Uhlenbeck proposed that the observed magnetic moment was attributable to spin angular momentum.

The electron-spin Hypothesis not only provided an explanation for the observed magnetic moment, but also accounted for many other effects in atomic spectroscopy, including changes in spectral lines in the presence of a magnetic field (Zeeman effect), doublet lines in alkali spectra, and fine structure (close doublets and triplets) in the hydrogen spectrum.

The Dirac equation also predicted additional states of the electron that had not yet been observed.

Experimental confirmation was provided in 1932 by discovering the positron by the American physicist Carl David Anderson. Every particle described by the Dirac equation has to have a corresponding antiparticle, which differs only in charge. The positron is just such an antiparticle of the negatively charged electron having the same mass as the latter but a positive charge.

3.10 Identical Particles and Multielectron Atoms

Because electrons are identical to (i.e., indistinguishable from) each other, an atom's wave function with more than one electron must satisfy special conditions. The problem of identical particles does not arise in classical physics, where the objects are large-scale and can always be distinguished, at least in principle. However, there is no way to differentiate two electrons in the same atom, and the form of the wave function must reflect this fact. The overall wave function Ψ of a system of identical particles depends on all the particles' coordinates. If the coordinates of two of the particles are interchanged, the wave function must remain unaltered or, at most, undergo a change of sign; the change of sign is permitted because it is $\Psi 2$ that occurs in the physical interpretation of the wave function. If the sign of Ψ remains unchanged, the wave function is symmetric concerning interchange; if the sign changes, the function is antisymmetric.

The symmetry of the wave function for identical particles is closely related to the spin of the particles. In quantum field theory (see below Quantum electrodynamics), it can be shown that particles with half-integral spin (1/2, 3/2, etc.) have antisymmetric wave functions. They are called fermions after the Italian-born physicist Enrico Fermi. Examples of fermions are electrons, protons, and neutrons, all of which have spin 1/2. Particles with zero or integral spin (e.g., mesons, photons) have symmetric wave functions. They are called bosons after the Indian mathematician and physicist Satyendra Nath Bose who first applied the ideas of symmetry to photons in 1924–25.

The requirement of antisymmetric wave functions for fermions leads to a fundamental result, known as the exclusion principle, first proposed in 1925 by the Austrian physicist Wolfgang Pauli. The exclusion principle states that two fermions in the same system cannot be in the same quantum state. If they were, interchanging the two sets of coordinates would not change the wave function, which contradicts the result that the wave function must change sign. Thus, two electrons in the same atom cannot have identical values for the four quantum numbers n, l, m, ms. The exclusion principle forms the basis of many properties of matter, including the periodic classification of the elements, the nature of chemical bonds, and the behavior of electrons in solids; the last determines in turn whether a solid is a metal, an insulator, or a semiconductor (see atom; matter).

The Schrödinger equation cannot be solved precisely for atoms with more than one electron. The principles of the calculation are well understood, but the problems are complicated by the number of particles and the variety of forces involved. The forces include the electrostatic forces between the nucleus and the electrons, and between the electrons themselves and weaker magnetic forces arising from the spin and orbital motions of the electrons. Despite these difficulties, approximation methods introduced by the English physicist Douglas R. Hartree, the Russian physicist Vladimir Fock, and others in the 1920s and 1930s, have achieved considerable success. Such schemes start by assuming that each electron moves independently in an average electric field because of the nucleus and the other electrons; i.e., correlations between the positions of the electrons are ignored. Each electron has its own wave function, called an orbital. The overall wave function for all the electrons in the atom satisfies the exclusion principle. Corrections to the calculated energies are then made, which depend on the strengths of the electron-electron correlations and the magnetic forces.

3.11 Tunneling

The quantum mechanics phenomenon of tunneling, which has no analog in classical physics, is a successful outcome. Consider a particle in the inner zone of a one-dimensional functional well $V(x)$ with energy E. In classical mechanics, if $E < V_0$ (the maximum height of the potential barrier), the particle remains in the well forever; if $E > V_0$, the particle escapes. In quantum mechanics, the situation is not so simple. The particle can escape even if its energy E is below the height of the barrier V_0, although the probability of escape is small, unless E is close to

V0. In that case, the particle may tunnel through the potential barrier and emerge with the same energy E.

The phenomenon of tunneling has many important applications. For example, it describes a type of radioactive decay in which a nucleus emits an alpha particle (a helium nucleus). According to the quantum explanation given independently by George Gamow and by Ronald W. Gurney and Edward Condon in 1928, the alpha particle is confined before the decay by a potential of the shape shown in Figure 1. A given nuclear species can measure the energy E of the emitted alpha particle and the average lifetime τ of the nucleus before decay. The lifetime of the nucleus is a measure of the probability of tunneling through the barrier — the shorter the lifetime, the higher the probability. With plausible assumptions about the general form of the potential function, it is possible to calculate a relationship between τ and E that applies to all alpha emitters. This theory, which is borne out by experiment, shows that the probability of tunneling, hence the value of τ, is extremely sensitive to E's value. For all known alpha-particle emitters, the value of E varies from about 2 to 8 million electron volts, or MeV (1 MeV = 10^6 electron volts). Thus, the value of E varies only by a factor of 4, whereas the range of τ is from about 10^{11} years down to about 10^{-6} second, a factor of 10^{24}. It would be difficult to account for this sensitivity of τ to the value of E by any theory other than quantum mechanical tunneling.

3.12 Axiomatic Approach

Although the two Schrödinger equations form an important part of quantum mechanics, it is possible to present the subject in a more general way. Dirac gave an elegant exposition of an axiomatic approach based on observables and states in a classic textbook entitled The Principles of Quantum Mechanics. (The book, published in 1930, is still in print.) An observable is anything that can be measured — energy, position, a component of angular momentum, and so forth. Every observable has a set of states, each state is represented by an algebraic function. With each state is associated a number that gives the result of a measurement of the observable. Consider an observable with N states, denoted by $\psi_1, \psi_2, \ldots, \psi_N$, and corresponding measurement values $a_1, a_2, \ldots, a_N$. A physical system — e.g., an atom in a particular state — is represented by a wave function Ψ, which can be expressed as a linear combination, or mixture, of the states of the observable. Thus, the Ψ may be written especially in the article "Quantum Mechanics." For a given Ψ, the quantities c_1, c_2, etc., are a set of numbers that can be calculated. In general, the numbers are complex, but they are assumed to be real numbers in the present discussion.

The theory postulates that the result of measurement must be an a-value — i.e., a_1, a_2, or a_3, etc. No other value is possible. Second, before the measurement is made, the probability of obtaining the value a_1 is c_1^2, and that of obtaining the value a_2

is c_{22}, and so on. If the value obtained is, say, a_5, the theory asserts that after the measurement, the system's state is no longer the original Ψ but has changed to ψ_5, the state corresponding to a_5.

Several consequences follow from these assertions. First, the result of a measurement cannot be predicted with certainty. Only the probability of a particular result can be predicted, even though the initial state (represented by the function Ψ) is known exactly. Second, identical measurements made on a large number of identical systems, all in the identical state Ψ, will produce different values for the measurements. This is, of course, quite contrary to classical physics and common sense, which say that the same measurement on the same object in the same state must produce the same result. Moreover, according to the theory, the act of measurement changes the state of the system, but it does so in an indeterminate way. Sometimes it changes the state to ψ_1, sometimes to ψ_2, and so forth.

There is an important exception to the above statements. Suppose that, before the measurement is made, the state Ψ happens to be one of the ψs — say, $\Psi = \psi_3$. Then $c_3 = 1$, and all the other cs are zero. This means that, before the measurement is made, obtaining the value a_3 is unity, and the probability of obtaining any other value of a is zero. In other words, in this particular case, the measurement's result can be predicted with certainty.

Moreover, after the measurement is made, the state will be ψ_3; the same as before. Thus, in this particular case, measurement does not disturb the system. Whatever the initial state of the system, two measurements made in rapid succession (so that the change in the wave function given by the time-dependent Schrödinger equation is negligible) produce the same result.

The value of one observable can be determined by a single measurement. The value of two observables for a given system may be known at the same time, provided that the two observables have the same set of state functions $\psi_1, \psi_2, \ldots, \psi_N$. In this case, measuring the first observable results in a state function that is one of the ψs. Because this is also a state function of the second observable, the result of measuring the latter can be predicted with certainty. Thus, the values of both observables are known. (Although the ψs are the same for the two observables, the two sets of values are, in general, different.) The two observables can be measured repeatedly in any sequence. After the first measurement, none of the measurements disturbs the system, and a unique pair of values for the two observables is obtained.

3.13 Incompatible Observables

The measurement of two observables with different sets of state functions is quite a different situation. Measurement of one observable gives a certain result. The state function after the measurement is, as always, one of the states of that observable; however, it is not a state function for the second observable. Measuring the second observable disturbs the system, and the system's state is no longer one of the states of the first observable. In general, measuring the first observable again does not produce the same result as the first time. To sum up, both quantities cannot be known at the same time, and the two observables are said to be incompatible.

A specific example of this behavior is the measurement of the component of angular momentum, along with two mutually perpendicular directions. The Stern-Gerlach experiment mentioned above involved measuring the angular momentum of a silver atom in the ground state. In reconstructing this experiment, a beam of silver atoms is passed between the poles of a magnet. The poles are shaped so that the magnetic field varies greatly in strength over a very small distance. The apparatus determines the ms quantum number, which can be +1/2 or −1/2. No other values are obtained. Thus, in this case, the observable has only two states — i.e., N = 2. The inhomogeneous magnetic field produces a force on the silver atoms in a direction that depends on the atoms' spin state. A beam of silver atoms is passed through magnet A. The atoms in the state with ms = +1/2 are deflected upward and emerge as beam 1 while those with ms = −1/2 are deflected downward and emerge as beam 2. If the magnetic field's direction is the x-axis, the apparatus measures Sx, which is the x-component of spin angular momentum. The atoms in beam 1 have Sx = $+\hbar/2$ while those in beam 2 have Sx = $-\hbar/2$. In a classical picture, these two states represent atoms spinning about the direction of the x-axis with opposite senses of rotation.

The y-component of spin angular momentum Sy can have only the values $+\hbar/2$ and $-\hbar/2$; however, the two states of Sy are not the same as for Sx. In fact, each of the states of Sx is an equal mixture of the states for Sy and conversely. Again, the two Sy states may be pictured as representing atoms with opposite senses of rotation about the y-axis. These classical pictures of quantum states are helpful, but only up to a certain point. For example, quantum theory says that each of the states is corresponding to spin about the x-axis is a superposition of the two states with spin about the y-axis. There is no way to visualize this; it has absolutely no classical counterpart. One simply has to accept the result as a consequence of the axioms of the theory. Suppose that the atoms in beam 1 are passed into a second magnet B, which has a magnetic field along the y-axis perpendicular to x. The atoms emerge from B and go in equal numbers through its two output channels. The classical theory says that the two magnets together have measured both the x and y-components of spin angular momentum and that the atoms in beam 3 have Sx = $+\hbar/2$, Sy = $+\hbar/2$ while those in beam 4 have Sx = $+\hbar/2$, Sy = $-\hbar/2$. However, classical theory is wrong because if beam 3 is put through still another magnet C, with its magnetic field along x, the atoms divide equally into beams 5 and 6 instead of emerging as a single beam 5 (as they would if they had Sx = $+\hbar/2$). Thus, the correct statement is that the beam entering B has Sx = +$\hbar$/2 and is composed of an equal mixture of the states Sy = +$\hbar$/2 and Sy = −$\hbar$/2—i.e., the x-component of angular momentum is known, but the y-component is not. Correspondingly, beam 3 leaving B has Sy = +$\hbar$/2 and is an equal mixture of the states Sx = +$\hbar$/2 and Sx = −$\hbar$/2; the y-component of angular momentum is known, but the x-component is not. The information about Sx is lost because of the disturbance caused by magnet B in the measurement of Sy.

3.14 Heisenberg Uncertainty Principle

The observables discussed so far have had discrete sets of experimental values. For example, the values of a sound system's energy are always discrete. Angular momentum components have values that take the form m$\hbar$, where m is either an integer or a half-integer, positive or negative. On the other hand, a particle's position or the linear momentum of a free particle can take continuous values in both quantum and classical theory. The mathematics of observables with a continuous spectrum of measured values is somewhat more complicated than for the discrete case but presents no principle problems. An observable with a continuous spectrum of measured values has an infinite number of state functions. The state function Ψ of the system is still regarded as a combination of the observable state functions, but the sum in equation (10) must be replaced by an integral.

Measurements can be made of position x of a particle and the x-component of its linear momentum, denoted by px. These two observables are incompatible because they have different state functions. The phenomenon of diffraction noted above illustrates the impossibility of measuring position and momentum simultaneously and precisely. If a parallel monochromatic light beam passes through a slit, its intensity varies with direction. The light has zero intensity in certain directions. Wave theory shows that the first zero occurs at an angle θo, given by sin θo = λ/b, where λ is the wavelength of the light and b is the width of the slit. If the width of the slit is reduced, θo increases — i.e., the diffracted light is more spread out. Thus, θo measures the spread of the beam.

The experiment can be repeated with a stream of electrons instead of a beam of light. According to de Broglie, electrons have wave-like properties; therefore, the beam of electrons emerging from the slit should widen and spread like a beam of light waves. This has been observed in experiments. If the electrons have velocity u in the forward direction (i.e., the y-direction mentioned above), their (linear) momentum is p = meu. Consider px, the component of momentum in the x-direction. After the electrons have passed through the aperture, the spread in their directions results in an uncertainty in px by a special amount composition for the article "Quantum Mechanics" where λ is the wavelength of the electrons according to the de Broglie formula, equals h/p. Thus, Δpx ≈

h/b. Exactly where an electron passed through the slit is unknown; it is certain that an electron went through somewhere. Therefore, immediately after an electron goes through, the uncertainty in its x-position is $\Delta x \approx b/2$. Thus, the product of the uncertainties is of the order of $\hbar$. The more exact analysis shows that the product has a lower limit, given by special composition for the article "Quantum Mechanics:" Heisenberg uncertainty principle.

This is the well-known Heisenberg uncertainty principle for position and momentum. It states that there is a limit to the precision with which the position and the momentum of an object can be measured simultaneously. Depending on the experimental conditions, either quantity can be measured as precisely desired (at least in principle). Still, the more precisely one of the quantities is measured, the less precisely the other is known.

The uncertainty principle is significant only on the atomic scale because of the small value of h in everyday units. If the position of a macroscopic object with a mass of, say, one gram is measured with a precision of 10−6 meters, the uncertainty principle states that its velocity cannot be measured to better than about 10−25 meters per second. Such a limitation is hardly worrisome. However, if an electron is located in an atom about 10−10 meters across, the principle gives a minimum uncertainty in the velocity of about 106 meters per second.

The above reasoning leading to the uncertainty principle is based on the wave-particle duality of the electron. When Heisenberg first propounded the principle in 1927, his reasoning was based on the photon's wave-particle duality. He considered the process of measuring the position of an electron by observing it in a microscope. Diffraction effects due to the wave nature of light result in a blurring of the image; the resulting uncertainty in the electron's position is approximately equal to the wavelength of the light. To reduce this uncertainty, it is necessary to use the light of a shorter wavelength — e.g., gamma rays. However, in producing an image of the electron, the gamma-ray photon bounces off the electron, giving the Compton effect (see above Early developments: Scattering of X-rays). As a result of the collision, the electron recoils in a statistically random way. The resulting uncertainty in the momentum of the electron is proportional to the momentum of the photon, which is inversely proportional to the wavelength of the photon. So, again, the case that increased precision in the knowledge of the position of the electron is gained only at the expense of decreased precision in the knowledge of its momentum. A detailed calculation of the process yields the same result as before. Heisenberg's reasoning clearly reveals that the smaller the particle being observed, the more significant the uncertainty principle. When a large body is observed, photons still bounce off it and change its momentum, but considered a fraction of the body's initial momentum, the change is insignificant.

The Schrödinger and Dirac theories give a precise value for each stationary state's energy, but the states do not have precise energy in reality. The only exception is in the ground (lowest energy) state. Instead, the energies of the states are spread over a small range. The spread arises from the fact that because the electron can transition to another state, the initial state has a finite lifetime. The transition is a random process, and so different atoms in the same state have different lifetimes. Suppose the mean lifetime is denoted as τ. In that case, the theory shows that the initial state's energy has a spread of energy ΔE, given by special composition for the article "Quantum Mechanics."

This energy spread is manifested in a spread in the frequencies of emitted radiation. Therefore, the spectral lines are not infinitely sharp. (Some experimental factors can also broaden a line, but their effects can be reduced; however, the present effect, known as natural broadening, is fundamental and cannot be reduced.) Equation (13) is another type of Heisenberg uncertainty relation; generally, if a measurement with duration τ is made of the energy in a system, the measurement disturbs the system, causing the energy to be uncertain by an amount ΔE, the magnitude of which is given by the above equation.

3.15 Quantum Electrodynamics

The application of quantum theory to the interaction between electrons and radiation requires a quantum treatment of Maxwell's field equations, which are the foundations of electromagnetism. The relativistic theory of the electron is formulated by Dirac (see above Electron spin and antiparticles). The resulting quantum field theory is known as quantum electrodynamics or QED.

QED accounts for the behavior and interactions of electrons, positrons, and photons. It deals with processes involving the creation of material particles from electromagnetic energy and the converse processes in which a material particle and its antiparticle annihilate each other and produce energy. Initially, the theory was beset with formidable mathematical difficulties because the calculated values of quantities such as the charge and mass of the electron proved infinite. However, an ingenious set of techniques developed (in the late 1940s) by Hans Bethe, Julian S. Schwinger, Tomonaga Shin'ichirō, Richard P. Feynman, and others dealt systematically with the infinities to obtain finite values of the physical quantities. Their method is known as "renormalization." The theory has provided some remarkably accurate predictions.

According to the Dirac theory, two particular states in hydrogen with different quantum numbers have the same energy. QED, however, predicts a small difference in their energies; the difference may be determined by measuring the frequency of the electromagnetic radiation that produces transitions between the two states. This effect was first measured by Willis E. Lamb, Jr., and Robert Retherford in 1947. Its physical origin lies in the electron's interaction with the random fluctuations in the surrounding electromagnetic field. These fluctuations, which exist even in the absence of an applied field, are quantum phenomena. The accuracy of experiment and theory in this area may be gauged by two recent values for the separation of the two states, expressed in terms of the frequency of the radiation that produces the transitions.

An even more spectacular example of the success of QED is provided by the value for μe, the magnetic dipole moment of the free electron. Because the electron is spinning and has an electric charge, it behaves like a tiny magnet; the strength of which is expressed by the value of μe. According to the Dirac theory, μe is exactly equal to $\mu B = e\hbar/2me$, a quantity known as the Bohr magneton; however, QED predicts that $\mu e = (1 + a)\mu B$, where a is a small number, approximately 1/860. Again, the QED correction's physical origin is the interaction of the electron with random oscillations in the surrounding electromagnetic field. The best experimental determination of μe involves measuring the quantity itself and the small correction term $\mu e - \mu B$. This greatly enhances the sensitivity of the experiment.

3.16 Quantum Harmonic Oscillator

There are two sorts of states in quantum mechanics:

1. **Bound states:** the particle is somewhat localized and cannot escape the potential.

2. **Unbound states:** the particle can escape the potential.

Note that for the same potential, whether something is a bound state or an unbound state, depends on the energy considered.

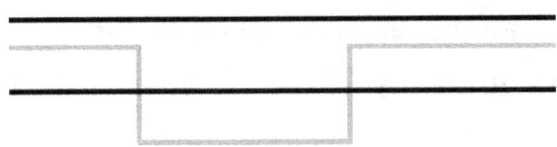

Figure 1: For the finite well, the energy represented by the lower black line is for a bound state, while the energy represented by the upper black line is for an unbound state.

But note that in quantum mechanics, because of the possibility of tunneling as seen before, the definition of whether a state is bound or not differs between classical and quantum mechanics. The point is that we need to compare E with $\lim_{x \to \pm\infty} V(x)$ to determine if a state is bound or not.

Why do we split our cases that way? Why do we study bound and unbound states separately if they obey the same equations? After all, in classical mechanics, both obey $F = \dot{p}$. In quantum mechanics, both obey:

$$\hat{E}\psi(x,t) = iJ\frac{\partial \psi(x,t)}{\partial t}$$

Figure 2: The same energy denoted by the black line is a bound classical and quantum state for the potential on the left, while the classical bound state is a quantum unbound state for the potential on the right.

The distinction is worth making because the bound and unbound states exhibit qualitatively different behaviors:

Mechanics	Bound States	Unbound States
Classical	Periodic motion	Aperiodic motion
Quantum	Discrete energy spectrum	Continuous energy spectrum

For now, we will focus on bound states, with discussions of unbound states coming later. Let us remind ourselves of some of the properties of bound states.

1. **Infinite Square Well**

- Ground state: no nodes
- Nth excited state: n nodes (by the node theorem)
- Energy eigenfunctions chosen to be real.
- Time evolution of energy eigenfunctions through a complex phase.
- Different states evolve at different rates.
- Energy eigenstates have no time evolution in observables as p(x) for such states is independent of t.
- Time evolution of expectation values for observables comes only through interference terms between energy eigenfunctions.

2. **Symmetric Finite Square Well**
   - Node theorem still holds
   - V (x) is symmetric
   - Leads to symmetry or antisymmetric of $\varphi(x; E)$
   - Antisymmetric $\varphi(x; E)$ are fine as $|\varphi(x; E)|^2$ is symmetric
   - Exponential tails in classically forbidden regions lead to discrete energy spectrum (picture of shooting for finite square well?)

3. **Asymmetric Finite Square Well**
   - Node theorem still holds
   - Asymmetry of potential breaks (anti)symmetry of eigenfunctions
   - Shallow versus deep parts of well
   - Deeper part of well → shorter λ as p = hλ−1, so particle travels faster there
   - Deeper part of well → lesser amplitude as particle spends less time there so the probability density is less there.

4. **Harmonic Oscillator**
   - Node theorem still holds
   - Many symmetries present
   - Evenly spaced discrete energy spectrum is very special!

3.17 Quantum Entanglement

Entanglement occurs in cases where you have limited information on the condition of two processes. For example, two objects that you'll call c-ons may form your systems. The "c" is supposed to suggest "classical," so you might think of your c-ons as cakes if you would like to have something unique and interesting in mind.

Your c-ons are available in two shapes which you define as their potential states. Then, for two c-ons, the four feasible joint states are (circle, circle) (square, square), (square, circle), and (circle, square). In each of those four states, the following tables show two instances of the possibilities for locating the system in all states.

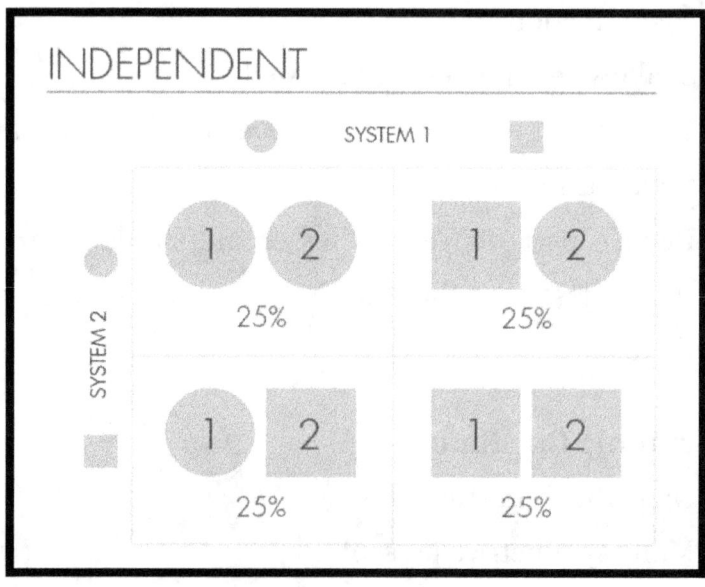

If information of the state of one of them does not provide useful data on the state of the other, you claim that the c-ons are "independent." This property is explained in the first table. If the first c-ons (or cake) are square, so you still do not know any information regarding the second's form. Likewise, the second's form would not disclose something useful about the first's shape.

On the other side, as details regarding one strengthen your perception of the other, you claim your two c-ons are intertwined. Extreme entanglement is illustrated in your second table. In that case, you realize the second cake is circular, too, whenever the first c-ons are circular. And whenever square is the first c-on, so too is the second. When you know the structure of one, you may confidently assume the type of the other.

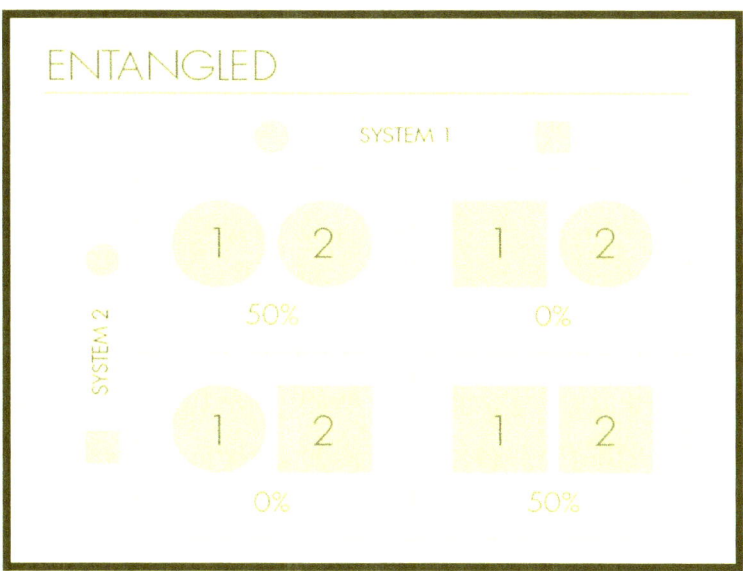

The quantum form of entanglement is the same phenomenon that lacks independence. In quantum theory, mathematical objects are used to describe states known as wave functions. Very interesting complications are introduced by the instructions connecting wave functions to physical probabilities. But the core principle of intertwined information, as for classical probabilities you have already discussed, carries over.

Of course, cakes cannot be counted as quantum systems. Still, entanglement among quantum systems occurs normally in the aftermath of particle collisions. One practical work, unentangled (independent) states, is an unusual exception as interaction generates associations between them whenever structures communicate.

They are considering molecules as an example. Subsystems include composites, including electrons and nuclei, the lost energy condition of a molecule, in which it is most frequently observed, is the strongly entangled state of its nuclei and electrons since the locations of certain constituent particles cannot be isolated in anyways. The electrons shift with them while the nuclei move.

Explaining the above examples: explaining square or circular states of system 1, it is written as Φ■, Φ• for the wave functions. While for system 2, it is written as ψ■, ψ• for the same states. Then in this working example, the whole state will be:

- Entangled: Φ■ ψ■ + Φ• ψ•
- Independent: Φ■ ψ■ + Φ■ ψ• + Φ• ψ■ + Φ• ψ

Independent version can be written as:
(Φ■ + Φ•)(ψ■ + ψ•)

Note: in that way in this design, the parentheses distinct systems 1 & 2 in the independent parts.

There are several ways that entangled states can be formed. One approach is to calculate the composite system that is providing partial results. You will be explained through example that two structures have conspired to have the same structure without knowing precisely what form they have. Later, this notion would become important.

The more distinctive implications of the quantum entanglement, such as the Greenberger Horne Zeilinger (GHZ) impacts and Einstein-Podolsky-Rosen (EPR), emerge from its relationship with another component of the quantum theory named "complementarity." Complementarity is required to be introduced for discussion of EPR and GHz.

Previously, you figured that your C-ons should represent two shapes (square and circle). You now imagine the two colors, red and blue, will also be exhibited along with shapes. Talking about classical structures, this added property would mean that your c-ons might be in either of four potential states. If you're talking about classical structures, possible results will include a red circle, red square, blue circle, or blue square.

However, with a quantum cake, maybe a quake, or a q-on (with more dignity), the case is radically different. A q-on will show various shapes of different colors in different conditions and does not imply that it simultaneously has both a form and a color. As you'll see soon, the "common sense" inference that Einstein believed should be part of every reasonable definition of physical reality is inconsistent with experimental evidence.

The shape of q-on can be measured, but it will lead to the loss of all information. Similarly, the color of q-on can also be measured, but it will be losing all information about its shape of q-on. According to quantum theory, it is not possible to measure both its color and shape simultaneously. One view of physical reality never explains all its aspects at once. As Niels Bohr framed it, it is the core of complementarity.

As a result, in assigning observable existence to individual properties, quantum theory forces one to be circumspect. It would help if you accepted the following to prevent contradictions.

1. A property that cannot be measured needs no existence.
2. Measurement is a vigorous process that changes the system that is being measured.

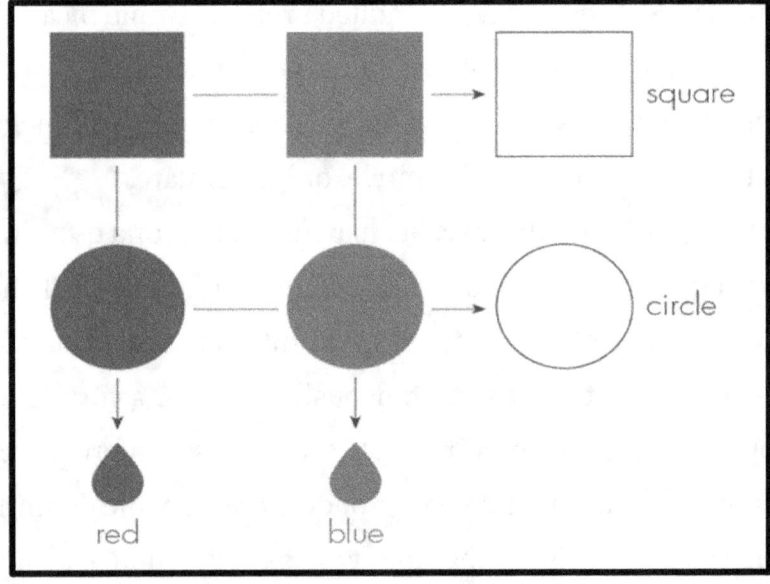

Two classics will be discussed here now, which are far from classical and illustrations of quantum theory's strangeness. In rigorous tests, these have been confirmed. (In real experiments, people calculate properties such as electron's angular momentum rather than cakes' shapes or colors.

A surprising effect of two quantum systems is Albert Einstein, Nathan Rosen (EPR), and Boris Podolsky. The EPR effect joins with an experimentally realizable type of quantum entanglement along with complementarity.

There are two q-ons in an EPR pair, one of which can be calculated either for its form or color (but not possible for both). You presume that several such pairs are all similar and that you may select which measurements to create from their components. You notice it is equally probable to be square or circular if you calculate the form of one element of an EPR pair. You notice it is equally probable to be red or blue whether you calculate the color.

The fascinating effects, which EPR proved paradoxical, originates when your measurements of all pair members are taken. When you measure both members for shape and colors, you discover that the consequences always agree. Consequently, if you find that one is red and later check the other's color, you will discover that it will be red too, and so forth. Besides this, if you measure the shape of one and later the other's color, there will be no correlation. Therefore, if the first comes in the square, then the second is correspondingly red or blue.

According to quantum theory, even though great distances divide the two systems, you can get certain effects, and the calculations are done almost simultaneously. In one location, the selection of measurement tends to influence the system's status in the other location. As Einstein named it, this "spooky action at a distance" may appear to entail the transfer of knowledge at a rate greater than the speed of light; in this case, details on what calculation was done.

But is it? You do not know what to think until you obtain the results. Only when you learn the effect you have calculated, valuable knowledge is obtained, not when it is calculated. The result revealing message needs to be transmitted in concrete physical ways, presumably slow you are compared to light speed.

The paradox disintegrates further upon further thought. Let us consider the second system's condition again, as the first was calculated to be red. If you want to test the color of the second q-on, you will get red. But when adding complementarity, as you mentioned earlier, if you want to calculate the form of a q-on, when it is in the "red" condition, you will have an equal chance of having a square or a circle. Therefore, the EPR result is theoretically induced, far from implementing a paradox. It is a repackaging of complementarity.

Neither is it paradoxical to find that there is a connection between distant events. And besides, if a pair of gloves are placed in boxes for each member and mail them to opposite directions of the world, one should not be shocked that one will evaluate the glove's handiness in the other from looking inside one package. Similarly, when the participants are close together, the correlations of an EPR pair must be imprinted.

Although in all established circumstances, they will withstand subsequent separation as they had memories. Again, there is no connection with a peculiarity of EPR as such, but its potential expression is complementary.

Daniel Greenberger, Anton Zeilinger, and Michael Horne found another beautifully illuminating example of quantum entanglement. Three of your q-ons are used; prepared in a unique, entangled condition (GHZ state). The three q-ons are being spread to three remote experimenters. Each experimenter selects whether to calculate form or color, individually or at random and notes the outcome. The procedure is replicated several times, always beginning with the GHZ Staten case of all three q-ons.

Each experimenter finds sufficiently random outcomes independently. She is equally capable of finding a square or a circle when she determines a q-on's shape. Similarly, when she calculates its color, red or blue are equally likely. So far, so earthly.

But later, a bit of study shows a stunning finding as the experimenters step together and discuss their measurements. Let us use the label "good" for square shapes and red colors and "evil" for circle shapes and blue colors. Experimenters noticed that when two of them wanted to measure shape but the third measured color, they realized that "evil" was exactly 0 or 2 effects (circular or blue). But they learned that precisely 1 or 3 measures You're evil as all three decided to quantify hue. It is what is expected by quantum mechanics, and that is what is observed.

So, is the evil accounts even or odd? Both possibilities are realized in various kinds of measurements with confidence. You are obliged to ignore the issue. Speaking of the sum of bad in your system, regardless of how it is calculated, makes little sense. Sometimes, it contributes to inconsistencies.

The GHZ effect is explained by "quantum mechanics in your face," in the scientist Sidney Coleman's words. It demolishes a profoundly established belief entrenched in everyday practice. It explains that physical systems have definite properties, regardless of whether such properties are calculated. For if they did, then measurement options would not impact the equilibrium of good and evil. The message of the GHZ influence is unforgettable and mind-blowing, once internalized.

So far, we have discussed how entanglement will make it difficult to assign many q-ons to a delegate's unique, autonomous state. Similar contributions can be applied to a single q-on in time for its evolution.

You claim. You have "entangled histories" because it is difficult to attribute a definitive status to your framework at any point in time. You may produce entangled history by creating measurements that collect partial knowledge about what occurred, like how You got traditional entanglement by removing certain possibilities. You have one q-on in the simplest embedded stories, which you track at two separate periods. You can visualize scenarios where you conclude that either your q-on form was always square or round at both times So that all alternatives are left in play by your observations. This is the simplest case of entanglement of quantum temporal analog already explained above.

Using a somewhat more elaborate procedure, you may apply the wrinkle of complementarity to this framework. You can describe conditions that bring out the "many worlds" feature of quantum theory. Thus, at an earlier point, your q-on could be prepared in the red state, and later it will be tested to be in the blue state. As in the basic examples above, you do not reliably apply the color property to your q-on, nor does it have a certain form. Histories of this type recognize the intuition that underlies the image of quantum mechanics in many worlds in a small yet, regulated and precise way. A definite state will branch into historical trajectories that are mutually inconsistent, which later come together.

Erwin Schrödinger was a founder and a pioneer of quantum theory. He was profoundly skeptical of the quantum theory's validity; emphasized that the evolution of quantum structures inevitably leads to states that may be tested with somewhat different properties. His "Schrödinger cat" notes, famously, explain quantum uncertainty is escalating into feline mortality problems. Before calculation, as you've demonstrated in your illustrations, one does not allocate the property of existence (or death) to the cat. Inside a netherworld of probability, both or neither coexists.

Everyday language is poorly adapted for describing quantum complementarity, partially because it is not experienced in everyday practice. Practical cats deal with certain air molecules in many different forms, among other items, based on whether they are alive or dead. Still, the calculation is performed immediately in nature, and the cat goes on with its existence (or death). Yet intertwined histories define q-ons that are Schrödinger kittens in a real sense. Their complete explanation includes that you take into consideration two contrasting property trajectories at intermediate times.

It is delicate to monitor the experimental realization of intertwined histories since it demands that you obtain partial data regarding your q-on. Conventional quantum measurements usually collect full details at one moment, such as specifying a definite shape or a definite hue, rather than many times spanning partial data. However, without considerable technological complexity, it can be achieved. In this sense, in quantum theory, you may give definite mathematical and experimental significance to the proliferation of "many worlds" and prove its substantiality.

CHAPTER 4:

The Interpretation of Quantum Mechanics

While quantum mechanics has been successfully applied to physics' problems, some of its concepts are strange. A few of their aspects are discussed here.

4.1 The Electron - Wave or Particle?

Young's experiment in which a parallel beam of monochromatic light is passed through a pair of narrow parallel slits has an electron counterpart. In Young's original experiment, the intensity of the light varies with direction after passing through the slits. The intensity oscillates because of interference between the light waves emerging from the two slits, the oscillation rate depending on the light's wavelength, and the separation of the slits. The oscillation creates a fringe pattern of alternating light and dark bands modulated by the diffraction pattern from each slit. If one of the slits is covered, the interference fringes disappear and only the diffraction pattern is left.

Young's experiment can be repeated with electrons, all with the same momentum. The screen in the optical experiment is replaced by a closely spaced grid of electron detectors. There are many devices for detecting electrons; the most common are scintillators. When an electron passes through a scintillating material, such as sodium iodide, the material produces a light flash that gives a voltage pulse amplified and recorded. The pattern of electrons recorded by each detector is the same as that predicted for waves with wavelengths given by the de Broglie formula. Thus, the experiment provides conclusive evidence for the wave behavior of electrons.

If the experiment is repeated with a very weak source of electrons so that only one electron passes through the slits, a single detector registers the arrival of an electron. This is a well-localized event characteristic of a particle. Each time the experiment is repeated, one electron passes through the slits and is detected. A graph plotted with detector position along one axis, and the number of electrons along the other looks exactly like the oscillating interference pattern. Thus, the intensity function in the figure is proportional to the probability of the electron moving in a particular direction after it has passed through the slits. Apart from its units, the function is identical to $\Psi 2$, where Ψ is the solution of the time-independent Schrödinger equation for this particular experiment.

If one of the slits is covered, the fringe pattern disappears and is replaced by the diffraction pattern for a single slit. Thus, both slits are needed to produce the fringe pattern. However, if the electron is a particle, it seems reasonable to suppose that it passed through only one of the slits. The apparatus can be modified to ascertain which slit by placing a thin wire loop around each slit. When an electron passes through a loop, it generates a small electric signal, showing which slit it passed through. However, the interference fringe pattern then disappears, and the single-slit diffraction pattern returns. Since both slits are needed for the interference pattern to appear and since it is impossible to know which slit the electron passed through without destroying that pattern, one is forced to conclude that the electron goes through both slits at the same time. In summary, the experiment shows both the wave and particle properties of the electron. The wave property predicts the probability of direction of travel before the electron is detected; on the other hand, the electron is detected in a particular place, showing that it has particle properties. Therefore, the answer to whether the electron is a wave or a particle is neither. It is an object exhibiting either wave or particle properties, depending on the type of measurement made on it. In other words, one cannot talk about the intrinsic properties of an electron; instead, one must consider the properties of the electron and measuring apparatus together.

4.2 Hidden Variables

A fundamental concept in quantum mechanics is that of randomness or indeterminacy. In general, the theory predicts only the probability of a certain result. Consider the case of radioactivity. Imagine a box of atoms with identical nuclei that can undergo decay with the emission of an alpha particle. In a given time interval, a certain fraction will decay. The theory may tell precisely what that fraction will be, but it cannot predict which particular nuclei will decay. The theory asserts that all the nuclei are in an identical state at the beginning of the time interval, and that decay is a completely random process. Even in classical physics, many processes appear random. For example, one says that when a roulette wheel is spun, the ball will drop at random into one of the numbered compartments in the wheel. Based on this belief, the casino owner and the players give and accept identical odds against each number for each throw. However, the fact is that the winning number could be predicted if one noted the wheel's exact location when the croupier released the ball, the initial speed of the wheel, and various other physical parameters. It is only ignorance of the initial conditions and the difficulty of doing the calculations that make the outcome appear random. In quantum mechanics, on the other hand, randomness is asserted to be absolutely fundamental. The theory says that, though one nucleus decayed and the other did not, they were previously in an identical state.

Many eminent physicists, including Einstein, have not accepted this indeterminacy. They have rejected the notion that the nuclei were initially in an identical state. Instead, they postulated that there must be some other property — presently unknown but existing nonetheless — different for the two nuclei. This type of unknown property is termed a hidden variable; it would restore determinacy to physics if it existed. If the initial values of the hidden variables were known, it would be possible to predict which nuclei would decay. Such a theory would, of course, also have to account for the wealth of experimental data, which conventional quantum mechanics explains from a few simple assumptions. Attempts have been made by de Broglie, David Bohm, and others to construct theories based on hidden variables, but the theories are very complicated and contrived. For example, the electron would definitely have to go through only one slit in the two-slit experiment. To explain why that interference occurs only when the other slit is open, it is necessary to postulate a special force on the electron, which exists only when that slit is open. Such artificial additions make hidden variable theories unattractive, and there is little support for them among physicists.

The orthodox view of quantum mechanics — and the one adopted in the present article — is known as the Copenhagen interpretation because its main protagonist, Niels Bohr, worked in that city. The Copenhagen view of understanding the physical world stresses the importance of basing theory on what can be observed and measured experimentally. It, therefore, rejects the idea of hidden variables as quantities that cannot be measured. The Copenhagen view is that the indeterminacy observed in nature is fundamental and does not reflect an inadequacy in present scientific knowledge. Therefore, one should accept indeterminacy without trying to "explain" it and see what consequences come from it.

Attempts have been made to link the existence of free will with the indeterminacy of quantum mechanics, but it is difficult to see how this feature of the theory makes free will more plausible. On the contrary, free will presumably imply rational thought and decision, whereas the essence of indeterminism in quantum mechanics is that it is due to intrinsic randomness.

4.3 Paradox of Einstein, Podolsky, and Rosen

In 1935, Einstein and two other physicists in the United States, Boris Podolsky, and Nathan Rosen, analyzed a thought experiment to measure position and momentum in a pair of interacting systems. Employing conventional quantum mechanics, they obtained some startling results which led them to conclude that the theory does not give a complete description of physical reality. Their results, which are so peculiar as to seem paradoxical, are based on impeccable reasoning, but their conclusion that the theory is incomplete does not necessarily follow. Bohm simplified their experiment while retaining the central point of their reasoning; this discussion follows his account.

The proton, like the electron, has spin 1/2; thus, no matter what direction is chosen for measuring the component of its spin angular momentum, the values are always $+\hbar/2$ or $-\hbar/2$. (The present discussion relates only to spin angular momentum, and the word spin is omitted from now on.) It is possible to obtain a system consisting of a pair of protons in close proximity and with total angular momentum equal to zero. Thus, if the value of one of the components of angular momentum for one of the protons is $+\hbar/2$ along any selected direction, the value for the component in the same direction for the other particle must be $-\hbar/2$. Suppose the two protons move in opposite directions until they are far apart. The total angular momentum of the system remains zero. If the component of angular momentum along the same direction for each of the two particles is measured, the result is a pair of equal and opposite values. Therefore, after the quantity is measured for one of the protons, it can be predicted for the other proton; the second measurement is unnecessary. As previously noted, measuring a quantity changes the state of the system. Thus, if measuring Sx (the x-component of angular momentum) for proton 1 produces the value $+\hbar/2$, proton 1 after measurement corresponds to Sx = $+\hbar/2$, and the state of proton 2 corresponds to Sx = $-\hbar/2$. Any direction, however, can be chosen for measuring the component of angular momentum. Whichever direction is selected, the state of proton 1 after measurement corresponds to a definite component of angular momentum about that direction.

Furthermore, since proton 2 must have the opposite value for the same component, it follows that the measurement on proton 1 results in a definite state for proton 2 relative to the chosen direction, notwithstanding the fact that the two particles may be millions of kilometers apart and are not interacting with each other at the time. Einstein and his two collaborators thought that this conclusion was so obviously false that the quantum mechanical theory on which it was based must be incomplete. They concluded that the correct theory would contain some hidden variable feature that would restore the determinism of classical physics.

A comparison of how quantum theory and classical theory describe angular momentum for particle pairs illustrates the essential difference between the two outlooks. In both theories, if a system of two particles has a total angular momentum of zero, then the angular momenta of the two particles are equal and opposite. If the components of angular momentum are measured along the same direction, the two values are numerically equal, one positive and the other negative. Thus, if one component is measured, the other can be predicted. The crucial difference between the two theories is that, in classical physics, the system under investigation is assumed to have possessed the quantity being measured beforehand. The measurement does not disturb the system; it merely reveals the preexisting state. It may be noted that if a particle were actually to possess components of angular momentum before measurement, such quantities would constitute hidden variables.

Does nature behave as quantum mechanics predicts? The answer comes from measuring the components of angular momentum for the two protons along with different directions with an angle θ between them. A measurement on one proton can give only the result $+\hbar/2$ or $-\hbar/2$. The experiment consists of measuring correlations between the plus and minus values for pairs of protons with a fixed value of θ and then repeating the measurements for different values of θ. The interpretation of the results rests on an important theorem by the Irish-born physicist John Stewart Bell. Bell began by assuming the existence of some form of a hidden variable with a value that would determine whether the measured angular momentum gives a plus or minus result. He further assumed locality, namely, that measurement on one proton (i.e., the choice of the measurement direction) cannot affect the result of the measurement on the other proton. Both these assumptions agree with classical, commonsense ideas. He then showed quite generally that these two assumptions lead to a certain relationship, now known as Bell's inequality, for the correlation values mentioned above. Experiments have been conducted at several laboratories with photons instead of protons (the analysis is similar), and the results show fairly conclusively that Bell's inequality is violated. That is to say, the observed results agree with those of quantum mechanics. They cannot be accounted for by a hidden variable (or deterministic) theory based on the concept of locality. One is forced to conclude that the two protons are a correlated pair and that a measurement on one affects the state of both, no matter how far apart they are. This may strike one as highly peculiar, but such is the way nature appears to be.

It may be noted that the effect on the state of proton 2 following a measurement on proton 1 is believed to be instantaneous; the effect happens before a light signal initiated by the measuring event at proton 1 reaches proton 2. Alain Aspect and his coworkers in Paris demonstrated this result in 1982 with an ingenious experiment. The correlation between the two angular momentum was measured, within a very short time interval, by a high-frequency switching device. The interval was less than the time taken for a light signal to travel from one particle to the other at the two measurement positions. Einstein's special theory of relativity states that no message can travel with a speed greater than that of light. Thus, there is no way that the information concerning the direction of the measurement on the first proton could reach the second proton before the measurement was made on it.

4.4 Measurement in Quantum Mechanics

The way quantum mechanics treats the process of measurement has caused considerable debate. Schrödinger's time-dependent wave equation is an exact recipe for determining the way the wave function varies with time for a given physical system in a given physical environment. According to the Schrödinger equation, the wave function varies in a strictly determinate way. On the other hand, in the axiomatic approach to quantum mechanics described above, measurement changes the wave function abruptly and discontinuously. Before the measurement is made, the wave function Ψ is a mixture of the ψs. The measurement changes Ψ from a mixture of ψs to a single ψ. This change, brought about by measurement, is termed "the collapse or reduction" of the wave function. The collapse is a discontinuous change in Ψ; it is also unpredictable. Starting with the same Ψ represented by the right-hand side of equation (10), the end result can be any individual ψs.

The Schrödinger equation, which gives a smooth and predictable variation of Ψ, applies between the measurements. However, the measurement process itself cannot be described by the Schrödinger equation; it is somehow a thing apart. This appears unsatisfactory since the measurement is a physical process and ought to be the Schrödinger equation subject just like any other physical process.

The difficulty is related to the fact that quantum mechanics applies to microscopic systems containing one (or a few) electrons, protons, or photons. Measurements, however, are made with large-scale objects (e.g., detectors, amplifiers, and meters) in the macroscopic world, which obeys the laws of classical physics. Thus, another way of formulating the question of what happens in measurement is to ask how the microscopic quantum world relates and interacts with the macroscopic classical world. More narrowly, it can be asked: how and at what point in the measurement process does the wave function collapse? There are no satisfactory answers to these questions, although there are several schools of thought. One approach stresses a conscious observer's role in the measurement process and suggests that the wave function collapses when the observer reads the measuring instrument. Bringing the conscious mind into the measurement problem seems to raise more questions than it answers, however.

As discussed above, the Copenhagen interpretation of the measurement process is essentially pragmatic. It distinguishes between macroscopic quantum systems and macroscopic measuring instruments. The initial object or event — e.g., the passage of an electron, photon, or atom — triggers the classical measuring device into giving a reading; somewhere along the chain of events, the result of the measurement becomes fixed (i.e., the wave function collapses). This does not answer the basic question but says, in effect, not to worry about it. This is probably the view of most practicing physicists.

The third school of thought notes that an essential feature of the measuring process is irreversibility. This contrasts with the wave function's behavior when it varies according to the Schrödinger equation; in principle, any such variation in the wave function can be reversed by an appropriate experimental arrangement. However, once a classical measuring instrument has given a reading, the process is not reversible. It is possible that the key to the nature of the measurement process lies somewhere here. The Schrödinger equation is known to apply only to relatively simple systems. It is an enormous extrapolation to assume that the same equation applies to the large and complex system of a classical measuring device. It may be that the appropriate equation for such a system has features that produce irreversible effects (e.g., wave function collapse) which differ in kind from those for a simple system.

One may also mention the so-called many-worlds interpretation, proposed by Hugh Everett III in 1957, which suggests that when a measurement is made for a system in which the wave function is a mixture of states, the universe branches into a number of noninteracting universes. Each of the possible outcomes of the measurement occurs but in a different universe. Thus, if $S_x = 1/2$ is the result of a Stern-Gerlach measurement on a silver atom (see above Incompatible observables), there is another universe identical to ours in every way (including clones of people), except that the result of the measurement is $S_x = -1/2$. Although this fanciful model solves some measurement problems, it has few adherents among physicists.

Because the various ways of looking at the measurement process lead to the same experimental consequences, trying to distinguish between them on scientific grounds may be fruitless. One or another may be preferred on the grounds of plausibility, elegance, or economy of hypotheses, but these are matters of individual taste. Whether one day a satisfactory quantum theory of measurement will emerge, distinguished from the others by its verifiable predictions, remains an open question.

CHAPTER 5:

Planck's Constant in Action

Planck's constant is a pretty important part of modern physics, but it is also pretty confusing. Maybe you were wondering what it was while reading about the new kilogram definition. Well, never fear, because this chapter will tell you everything you need to know about Planck's constant.

In the late 1800s, physics was facing a crisis. Physicists were trying to model atomic vibrations, but they kept getting it wrong. All the physics they knew at that point said it should look a certain way, but reality looked completely different, and no one knew why.

The problem would be solved by a man named Max Planck. Previous physicists had assumed that atomic vibrations were continuous; that is, they could vibrate at any frequency. Planck decided to assume that atoms could only vibrate at certain frequencies that were whole number multiples of some base frequency which he called h. In other words, atoms could vibrate at the h frequency, or 2h, or 3h, but not 2.5h.

This was a really bizarre assumption, but it worked. It turns out that atoms (and a lot of other stuff) can only take certain specific values. In physics, we say that atomic vibrations are quantized. Planck's discovery would ignite a flurry of research in a new field of physics called quantum mechanics. That h constant that Planck discovered would eventually be called Planck's constant, and it literally puts the "quantum" in "quantum mechanics."

5.1 The Invisible World of the Ultrasmall

Planck and other physicists in the late 1800s and early 1900s were trying to understand the difference between classical mechanics — that is, the motion of bodies in the observable world around us described by Sir Isaac Newton in the late 1600s — and an invisible world of the ultrasmall, where energy behaves in some ways like a wave and in some ways like a particle, also known as a photon.

"In quantum mechanics, physics works different from our experiences in the macroscopic world," explains Stephan Schlamminger, a physicist for the National Institute of Standards and Technology, by email. As an explanation, he cites the example of a familiar harmonic oscillator, a child on a swing set.

"In classical mechanics, the child can be at any amplitude (height) on the swing's path," Schlamminger says. "The Energy that the system has is proportional to the square of the amplitude. Hence, the child can swing at any continuous range of energies from zero up to a certain point."

But when you get down to the level of quantum mechanics, things behave differently. "The amount of energy that an oscillator could have is discrete, like rungs on a ladder," Schlamminger says. "The energy levels are separated by h times f, where f is the frequency of the photon — a particle of light — and an electron would release or absorb to go from one energy level to another."

But Max Planck found something very different when he looked deeper, he explains in the email. "Energy is quantized, or it is discrete, meaning I can only add one sugar cube or two or three. Only a certain amount of energy is allowed."

Planck's constant defines the amount of energy that a photon can carry according to the frequency of the wave in which it travels.

Electromagnetic radiation and elementary particles "display intrinsically both particle and wave properties," explains Fred Cooper, an external professor at the Santa Fe Institute, an independent research center in New Mexico, by email. "The fundamental constant which connects these two aspects of these entities is Planck's constant. Electromagnetic energy cannot be transferred continuously but is transferred by discrete photons of light whose energy E is given by $E = hf$, where h is Planck's constant, and f is the frequency of the light."

5.2 A Slightly Changing Constant

One of the confusing things for nonscientists about Planck's constant is that the value assigned to it has changed by tiny amounts over time. Back in 1985, the accepted value was $h = $

6.626176×10^{-34} Joule-seconds. The current calculation, done in 2018, is h = $6.62607015 \times 10^{-34}$ Joule-seconds.

"While these fundamental constants are fixed in the fabric of the universe, we humans don't know their exact values," Schlamminger explains. "We have to build experiments to measure these fundamental constants to the best of humankind's ability. Our knowledge comes from a few experiments that were averaged to produce a mean value for the Planck constant."

To measure Planck's constant, scientists have used two different experiments — the Kibble balance and the X-ray crystal density (XRCD) method. Over time, they have developed a better understanding of how to get a more precise number. "When a new number is published, the experimenters put forward their best number, as well as their best calculation of the uncertainty in their measurement," Schlamminger says. "The true, but unknown value of the constant, should hopefully lie in the interval of plus/minus the uncertainty around the published number, with a certain statistical probability." At this point, "we are confident that the true value is not far off. The Kibble balance and the XRCD method are so different that it would be a major coincidence that both ways agree so well by chance."

That tiny imprecision in scientists' calculations is not a big deal in the scheme of things. But if Planck's constant were a significantly bigger or smaller number, "all the world around us would be completely different," explains Martin Fraas, an assistant professor in mathematics at Virginia Tech, by email. If the value of the constant was increased, for example, stable atoms might be many times bigger than stars.

The size of a kilogram, which came into force on May 20, 2019, as agreed upon by the International Bureau of Weights and Measures (whose French acronym is BIPM), is now based upon Planck's constant.

CHAPTER 6:

Applications of Quantum Physics

Quantum mechanics has been extremely influential in describing microscopic phenomena in all branches of physics, as previously mentioned. The three phenomena mentioned in this section serve as examples of the theory's core concepts.

6.1 Decay of the Kaon

The kaon (also called the K0 meson), discovered in 1947, is produced in high-energy collisions between nuclei and other particles. It has zero electric charges, and its mass is about one-half the mass of the proton. It is unstable and, once formed, rapidly decays into either 2 or 3 pi-mesons. The average lifetime of the kaon is about 10^{-10} seconds.

In spite of the fact that the kaon is uncharged, quantum theory predicts the existence of an antiparticle with the same mass, decay products, and average lifetime; the antiparticle is denoted by K0. During the early 1950s, several physicists questioned the justification for postulating the existence of two particles with similar properties. In 1955, however, Murray Gell-Mann and Abraham Pais made an interesting prediction about the decay of the kaon. Their reasoning provides an excellent illustration of the quantum mechanical axiom that the wave function Ψ can be a superposition of states; in this case, there are two states: the K0 and K̄0 mesons themselves. A K0 meson may be represented formally by writing the wave function as Ψ = K0; similarly, Ψ = K̄0 represents a K̄0 meson. From the two states, K0 and K̄0, the following two new states are constructed:

$$K_1 = \frac{(K^0 + \bar{K}^0)}{\sqrt{2}},$$

$$K_2 = \frac{(K^0 - \bar{K}^0)}{\sqrt{2}}.$$

From these two equations, it follows that:

$$K^0 = \frac{(K_1 + K_2)}{\sqrt{2}},$$

$$\bar{K}^0 = \frac{(K_1 - K_2)}{\sqrt{2}}.$$

The reason for defining the two states K_1 and K_2 is that, according to quantum theory, when the K0 decays, it does not do so as an isolated particle; instead, it combines with its antiparticle to form the states K_1 and K_2. The state K_1 (called the K-short [K0S]) decays into two pi-mesons with a very short lifetime (about 9 × 10−11 second), while K2 (called the K-long [K0L]) decays into three pi-mesons with a longer lifetime (about 5 × 10−8 second). The physical consequences of these results may be demonstrated in the following experiment. K0 particles are produced in a nuclear reaction at point A (Figure 1). They move to the right in the figure and start to decay. At point A, the wave function is Ψ = K0, which can be expressed as the sum of K1 and K2. As the particles move to the right, the K1 state begins to decay rapidly. If the particles reach point B in about 10−8 second, nearly all the K1 component has decayed, although hardly any of the K2 components have done so. Thus, at point B, the beam has changed from pure K0 to one of almost pure K2, which shows an equal mixture of K0 and K0. In other words, K0 particles appear in the beam simply because K1 and K2 decay at different rates. At point B, the beam enters a block of absorbing material. Both the K0 and K0 are absorbed by the nuclei in the block, but the K0 are absorbed more strongly. As a result, even though the beam is an equal mixture of K0 and K0 when it enters the absorber, it is almost pure K0 when it exits at point C. The beam thus begins and ends as K0.

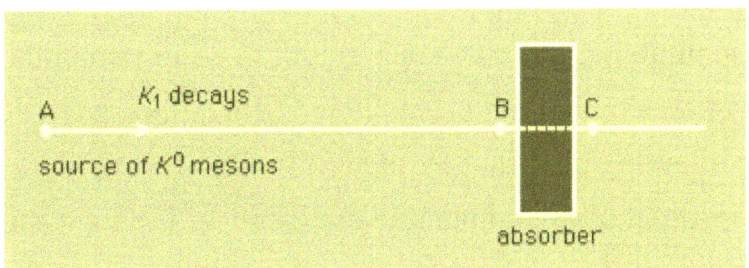

Figure 1: Decay of the K0 meson.

Gell-Mann and Pais predicted all this, and experiments subsequently verified it. The experimental observations are that the decay products are primarily two pi-mesons with a short decay time near A, three pi-mesons with longer decay time near B, and two pi-mesons again near C. (This account exaggerates the changes in the K1 and K2 components between A and B and in the K0 and K̄0 components between B and C; the argument, however, is unchanged.) The phenomenon of generating the K0 and regenerating the K1 decay is purely quantum. It rests on the quantum axiom of the superposition of states and has no classical counterpart.

6.2 Cesium Clock

The cesium clock is the most accurate type of clock yet developed. This device uses transitions between the spin states of the cesium nucleus and produces a frequency that is so regular that it has been adopted for establishing the time standard.

Like electrons, many atomic nuclei have spin. The spin of these nuclei produces small effects in the spectra known as hyperfine structure. (The effects are small because, though the angular momentum of a spinning nucleus is of the same magnitude as that of an electron, its magnetic moment, which governs the atomic levels' energies, is relatively small.) The nucleus of the cesium atom has spin quantum number $7/2$. The total angular momentum of the lowest energy states of the cesium atom is obtained by combining the nucleus's spin angular momentum with that of the single valence electron in the atom. (Only the valence electron contributes to the angular momentum because the angular momentum of all the other electrons total zero. Another simplifying feature is that the ground states have zero orbital momentum, so only spin angular momentum needs to be considered.) When the nuclear spin is considered, the atom's total angular momentum is characterized by a quantum number, conventionally denoted by F, which for cesium is 4 or 3. These values come from the spin value 7/2 for the nucleus and 1/2 for the electron. If the nucleus and the electron are visualized as tiny spinning tops, the value F = 4 (7/2 + 1/2) corresponds to the tops spinning in the same sense, and F = 3 (7/2 − 1/2) corresponds to spins in opposite senses. The energy difference ΔE of the states with the two F values is a precise quantity. If electromagnetic radiation of frequency v0 where special composition for the article "Quantum Mechanics" is applied to a cesium atoms system, transitions will occur between the two states. An apparatus that can detect the occurrence of transitions thus provides an extremely precise frequency standard. This is the principle of the cesium clock.

The apparatus is shown schematically in Figure 2. A beam of cesium atoms emerges from an oven at a temperature of about 100 °C. The atoms pass through an inhomogeneous magnet A, which deflects the atoms in state F = 4 downward and those in state F = 3 by an equal amount upward. The atoms pass through slit S and continue into a second inhomogeneous magnet B. Magnet B is arranged to deflect atoms with an unchanged state in the same direction that magnet A deflected them. The atoms follow the paths indicated by the broken lines in the figure and are lost to the beam. However, if an alternating electromagnetic field of frequency v0 is applied to the beam as it traverses the center region C, transitions between states will occur. Some atoms in state F = 4 will change to F = 3, and vice versa. For such atoms, the deflections in magnet B are reversed. The atoms follow the whole lines in the diagram and strike a tungsten wire which gives electric signals in proportion to the number of cesium atoms striking the wire. As the frequency v of the alternating field is varied, the signal has a sharp maximum for v = v0. The length of the apparatus from the oven to the tungsten detector is about one meter.

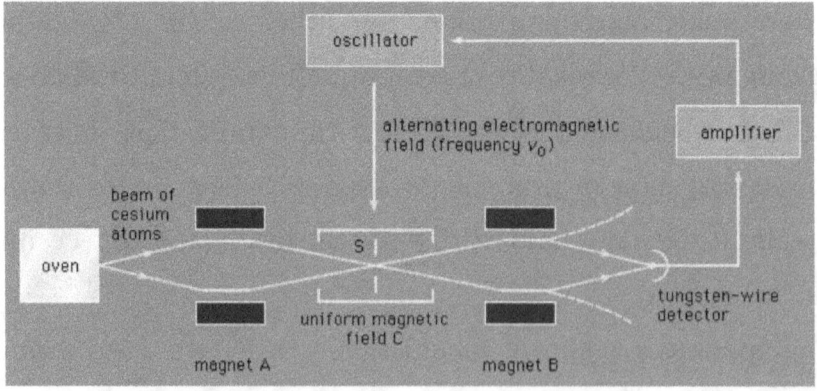

Figure 2: Cesium clock.

Each atomic state is characterized not only by the quantum number F, but also by a second quantum number mF. For F =

4, mF can take integral values from 4 to −4. In the absence of a magnetic field, these states have the same energy. However, a magnetic field causes a small change in energy proportional to the magnitude of the field and the mF value. Similarly, a magnetic field changes the energy for the F = 3 states according to the mF value, which, in this case, may vary from 3 to −3. The energy changes are indicated in Figure 3. A weak constant magnetic field is superposed on the alternating electromagnetic field in region C in the cesium clock. The theory shows that the alternating field can transition only between pairs of states with mF values that are the same or that differ by unity. However, as can be seen from the figure, the only transitions occurring at the frequency v0 are between the two states with mF = 0. The apparatus is so sensitive that it can discriminate easily between such transitions and all the others.

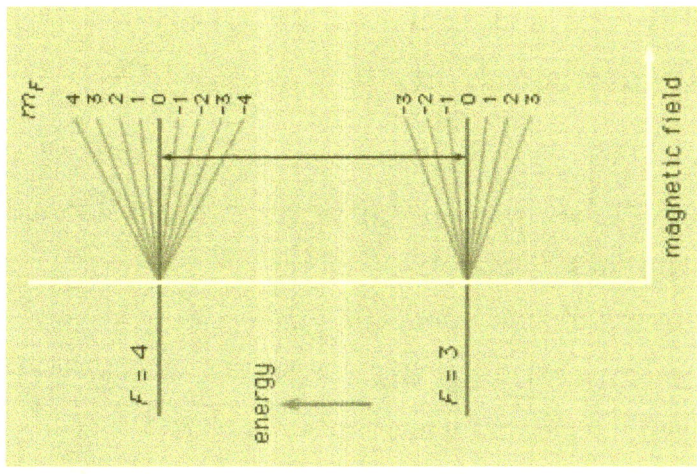

Figure 3: Variation of energy with magnetic-field strength for the F = 4 and F = 3 states in cesium-133.

If the frequency of the oscillator drifts slightly so that it does not quite equal v0, the detector output drops. The change in signal strength produces a signal to the oscillator to bring the frequency back to the correct value. This feedback system keeps the oscillator frequency automatically locked to v0.

The cesium clock is exceedingly stable. The frequency of the oscillator remains constant to about one part in 10^{13}. For this reason, the device is used to redefine the second. This base unit of time in the SI system is defined as equal to 9,192,631,770 cycles of the radiation corresponding to the transition between the levels $F = 4$, $mF = 0$ and $F = 3$, $mF = 0$ of the ground states of the cesium-133 atom. Before 1967, the second was defined in terms of the motion of Earth. The latter, however, is not nearly as stable as the cesium clock. Specifically, the fractional variation of Earth's rotation period is a few hundred times larger than that of the cesium clock frequency.

6.3 A Quantum Voltage Standard

Quantum theory has been used to establish a voltage standard, and this standard has proven to be extraordinarily accurate and consistent from laboratory to laboratory.

If two layers of superconducting material are separated by a thin insulating barrier, a supercurrent (i.e., a current of paired electrons) can pass from one superconductor to the other. This is another example of the tunneling process described earlier. Several effects based on this phenomenon were predicted in 1962 by the British physicist Brian D. Josephson.

Demonstrated experimentally soon afterward, they are now referred to as the Josephson effects.

If a DC (direct-current) voltage V is applied across the two superconductors, the energy of an electron pair changes by an amount of 2eV as it crosses the junction. As a result, the supercurrent oscillates with frequency ν given by the Planck relationship ($E = h\nu$).

This oscillatory behavior of the supercurrent is known as the "AC (alternating-current) Josephson effect." Measurement of V and ν permits a direct verification of the Planck relationship. Although the oscillating supercurrent has been detected directly, it is extremely weak. A more sensitive method of investigating is to study effects resulting from the interaction of microwave radiation with the supercurrent.

Several carefully conducted experiments have verified such a high degree of precision that it has been used to determine the value of 2e/h. This value can be determined more precisely by the AC Josephson effect than by any other method. The result is so reliable that laboratories now employ the AC Josephson effect to set a voltage standard.

In this way, measuring a frequency, which can be done with great precision, gives the value of the voltage. Before the Josephson method was used, the voltage standard in metrological laboratories devoted to the maintenance of physical units was based on high-stability Weston cadmium cells. These cells, however, tend to drift and so causing inconsistencies between standards in different laboratories. The Josephson method has provided a standard giving agreement within a few parts in 10^8 for measurements made at different times and in different laboratories.

The values of the fundamental constants, such as c, h, e, and me, are determined from a wide variety of experiments based on quantum phenomena. The results are so consistent that the values of the constants are thought to be known in most cases to better than one part in 10^8. Physicists may not know what they are doing when they make a measurement, but they do it extremely well.

Conclusion

These are difficult concepts for many people, even scientists, to grasp, and even Albert Einstein had significant philosophical issues with a universe that appears to behave in an utterly random manner at the subatomic level. There is a wealth of literature on the topic, both for beginners and experts. A few of these are listed in this book for a better or more detailed understanding of this complicated and confusing subject.

Despite its challenges, quantum theory remains an important part of modern physics' foundation. It is arguably one of the most successful theories in all of science. Despite its esoteric appearance, it is primarily a practical branch of physics; opening the way for applications such as the laser, electron microscope, transistor, superconductor, and nuclear power, as well as explaining important physical mechanisms such as chemical bonding, the configuration of the atom, and nuclear power.

Quantum theory, however, only effectively describes three of the four essential forces: the strong nuclear force, electromagnetism, and the weak nuclear force, despite its effectiveness in predicting and explaining the world around us. It does not clarify how gravity works. It attempts to merge quantum theory with the General Theory of Relativity in a single theory of quantum gravity, the so-called "theory of all," which is hoped to make sense of the entire universe, seems to be the way forward for physics.

www.ingramcontent.com/pod-product-compliance
Lightning Source LLC
Chambersburg PA
CBHW071521080526
44588CB00011B/1510